QUETZALS

Icons of the Cloud Forest

QUETZALS

Icons of the Cloud Forest

Alan F. Poole

**A Zona Tropical Publication
and Ojalá Ediciones**

Comstock Publishing Associates
an imprint of
Cornell University Press
Ithaca and London

First published 2023 by Cornell University Press

Printed in China

Library of Congress Cataloging-in-Publication Data

Names: Poole, Alan Forsyth, author.
Title: Quetzals : icons of the cloud forest / Alan F. Poole.
Description: Ithaca [New York] : Comstock Publishing Associates, an imprint of Cornell University Press, 2023. | Includes bibliographical references.
Identifiers: LCCN 2023002756 | ISBN 9781501772214 (paperback)
Subjects: LCSH: Quetzals. | Quetzals—Effect of human beings on—History. | Quetzals—Habitat—Conservation.
Classification: LCC QL696.T7 P66 2023 | DDC 598.7/3—dc23/eng/20230126
LC record available at https://lccn.loc.gov/2023002756

Zona Tropical Press ISBN 978-1-949469-40-0

Book design: Gabriela Wattson

Front cover: Male Resplendent Quetzal (Ondrej Prosicky / Shutterstock)

CONTENTS

PREFACE

Possibly no other feathered being of this hemisphere … has a longer history, as the philologist rather than the naturalist would use the term. This history is largely unwritten; and it is to be hoped that before long one who is at once an archaeologist and an ornithologist will make good the deficiency.

Alexander Skutch, "Life History of the Quetzal"

About 75 years ago, four years before I was born, Skutch laid out a challenge for some future author in the text above. It was, of course, the Resplendent Quetzal that this Costa Rican naturalist had in mind; other quetzals live well to the south, beyond what was his zone of study. Just a few years ago, and much to my surprise, I discovered that Skutch's challenge had yet to be answered, at least in any substantial way. Despite some misgivings, I decided to take the plunge.

By no means an archaeologist—and a philologist on only the rarest of days—I nonetheless felt that I brought some credibility to the task. The ornithologist's hat fit fairly comfortably; I had edited a publication on the life histories of North American birds for almost 25 years (although that left my knowledge of Neotropical birds a little weak). I had an abiding interest in avian life histories—the full portrait of a bird, something championed by Skutch, whose writing I had admired for years. And perhaps most importantly, in retirement I found myself living part of each year in the highlands of southern Costa Rica, on the edge of splendid quetzal habitat. Inspired by days in those magical forests, surrounded by courting and nesting Resplendents, a book project became tough to resist. And besides, I rationalized, who else was going to do it? At the time I was ready to step into this, no one was actively involved in research on any of the quetzals.

I soon discovered that someone *had* written a quetzal "history," at least a slice of it. In the early 1980s, J. E. Maslow set out to write what he described as a "political ornithology," a look at a politically and ecologically disintegrating country (Guatemala) through the lens of two very different birds (the Resplendent Quetzal and the Black Vulture, in *Bird of Life, Bird of Death*). Although a brilliant piece of writing, with on-target ornithological details, its focus was on the human element, and he provided a picture of human suffering at an especially grim moment in that country's history.

One of the aims of this book is to delve deep into the life history of the Resplendent Quetzal throughout its range—Mexico to Panama. Maslow ended his book at a then small and dysfunctional quetzal refuge that he predicted would soon be gone. Although problems remain there, the refuge and surrounding areas now host a number of thriving quetzal lodges, catering to a steady stream of tourists who have little trouble finding Resplendents in nearby forests. If nothing else, the quetzal story today needs to balance past pessimism with current realities.

Heading south from Guatemala today, for example, we find a more robust quetzal tourism in Costa Rica and Panama, thanks to significant cloud forest habitat set aside in private and government-protected reserves. All this reflects intense interest in the species and brings much needed revenue to communities, helping to spur habitat protection. Overall, estimates give us impressive numbers for the Resplendent Quetzal today: at least three-quarters of a million hectares of cloud forest habitat remaining throughout its range, with at least one nesting pair every 2 to 3 hectares, and sometimes more. For the moment, these numbers provide reason for optimism.

In addition to bringing the Resplendent record up to date, there are other reasons a quetzal book seems especially timely now. For starters, the Resplendent is not the only quetzal, and although its four *Pharomachrus* relatives remain little studied in their South American haunts, they nonetheless give us key reference points, especially evolutionarily, as we dive into the history of the better-known Resplendent. In addition, research has shown us fascinating ecological details for Resplendents, a bird whose diet centers on a dozen or so key fruits, and whose nesting depends on finding adequate numbers of dead trees with holes to accommodate eggs and young, likely a scarce resource.

What's more, Resplendents, and for that matter the other quetzals, open a fascinating window onto the dynamics of evolution. All are birds that evolution has pushed to extremes, especially in their dazzling plumage. Teasing out the selective forces that have nudged such extraordinary plumage into being provides some challenges. So far, no book has taken those up in detail; this book does.

The quetzal history that Skutch wanted to see written necessarily involves a look at early human civilizations in the Americas. For Aztec and Mayan civilizations, a thousand years ago (and beyond), Resplendent Quetzal feathers figured prominently in their religion; tens of thousands of these feathers were brought annually in trade to population centers, with unknown impacts on Resplendent populations in source areas. Recent research has shed some light on this fascinating trade, and how the feathers were procured. But, until now, no quetzal history has tried to tie the pieces together, especially with respect to how nesting populations might have held up under such an onslaught.

Above all, a quetzal history must look ahead to the conservation challenges these birds are likely to face over the next several decades. While we can take heart at the robust populations of Resplendents found in many protected reserves today, there is plenty to worry about on the horizon. Burgeoning human populations threaten forests in many of the highland belts that quetzals favor, even in some areas now considered "protected." Cattle and coffee farms bring major challenges, owing to their economic clout, but the damming of rivers for hydropower is a looming threat too, as is lumbering. Looking beyond these, the impacts of climate change—fast-rising temperatures and drying forests—pose by far the most worrisome challenges to quetzals. Elevations that currently harbor these birds are particularly vulnerable to such change, as the forest trees that feed and shelter quetzals are forced to adapt and shift range more quickly than previously.

Last, this book endeavors to celebrate these extraordinary creatures. I fervently hope I've succeeded in doing that. In quetzals, evolution has given us supremely dazzling birds, spun from an alchemy of sunlight and fruit in forests that harbor some of the richest biodiversity on the planet. Quetzals, especially the Resplendent, are increasingly the "flagship" species that help define these precious Mesoamerican cloud forests. But flags can rally people only when they're front and center, flying boldly. Here's hoping this book will bring the glory and fascination of quetzals to a host of people who have never known them before.

ACKNOWLEDGMENTS

For the people who helped make this book a reality, my gratitude runs deep and in many directions. Henry Barrantes of Sur Trips (San Vito, Costa Rica) introduced me to quetzals, driving me up steep, tortuous mountain roads in his capable 4-wheelers to the Las Tablas Protected Zone (Zona Protectora Las Tablas), in Costa Rica's southernmost Talamanca Mountains, where Resplendents reign supreme. The idea for this book would never have taken root without Henry as my enthusiastic guide, and without the solitude and splendor of highland Las Tablas and the bounty of quetzals there, so easily viewed. An added bonus was the chance to linger there for days at a time, taking in the birds and weather and landscapes, along with local history and culture, thanks to accommodations at Don Miguel Santi's farmhouse. There is good news: this area, portions of which are increasingly degraded and have long been held in private hands, is now starting to get the protection it deserves, owing to efforts by the Costa Rican government's National System of Conservation Areas (SINAC).

For help with research, especially the chance to learn about Resplendent Quetzals in regions I was unable to visit (COVID-19 trimmed my travel wings), I am indebted to Dr. Sofía Solórzano of Tlalnepantla, Mexico (UNAM). Her impressive research speaks for itself and provided an important leg up in my understanding of Resplendents in Chiapas, Mexico, and elsewhere. Particularly helpful were her papers on quetzal taxonomy, the first and only work on that topic, and her unpublished data on breeding density, which helped me flesh out population estimates across the Resplendent's range.

Knut Eisenmann (Cayaya Birding, Guatemala) helped me understand Resplendents in that country. A ready correspondent, I extend sincere thanks to him.

As even a quick glance at this book will show, Al Gilbert's splendid paintings of quetzals are an ornithological treasure. In granting me rights to his brushwork, Al boosted my confidence and earned my humble thanks. I can only hope that he will be as pleased as I am to see new readers opening pages to find his incomparable quetzals coming to life before their eyes.

Time, strength, and patience came to me from a variety of sources, I am pleased so say, but especially from the two homes that are dear to my heart, the Las Cruces Biological Station in San Vito, Costa Rica, from January to March, and, in the remainder of the year, the coast of southeastern Massachusetts, US. I am especially grateful to the staff at Las Cruces, who welcomed a stranger from afar, one with a poor grip on Spanish, and made him feel like family.

I was humbled and delighted by how many people answered my call for financial support when I first set out to make this book a reality. Major gifts arrived from Chris Baldwin; the Rev. Mark K. J. and Eleanor P. Robinson; Rob Bierregaard and Cathy Dolan; Luna Lodge (Osa Peninsula, Costa Rica); K. and K. Battle; R. Pasquier; R. Grew; M. and A. Copp; Georgia Chafee and John Nassikas; the Wagner Family and Weatherlow Foundation; and Ian Nisbet. In addition, numerous other donors stepped up to contribute, and I am no less grateful for their faith in my efforts. I can only hope that all will be as pleased with this book as I am.

Special thanks to Osa Conservation (OC), which has made significant steps in protecting and understanding the landscapes and ecology of Costa Rica's southern Pacific slope. OC served as my fiscal agent during the research and writing of this book, a very valuable contribution indeed.

Reviewers answered my call when I needed them most, substantially improving both my writing and the accuracy of this book. I remain particularly indebted to Roger Pasquier and Julie Zickefoose, who read the entire manuscript and made countless helpful suggestions, both sharpening my writing and getting me to rethink ideas that hadn't been fully developed. Nigel Collar (Birdlife International) and Mary C. Stoddard and David Ocampo (Princeton University) each edited the second chapter with a careful eye, helping to make it more accurate and coherent. In reading an early draft of the third chapter, Karl Taube (UC Riverside) brought his considerable insights on Mayan and Aztec civilizations to bear. I feel lucky to have had his help. In addition, Michael Coe (Yale University), a giant among Mayan scholars, let me invite myself for a visit to his New Haven home, just months before he died. To everyone, heartfelt thanks. Remaining errors are mine alone.

INTRODUCTION

… a bird of such incredible beauty that for 200 years European naturalists thought it must be the fabrication of American aboriginies [sic]. A bird so sacred to the ancient Maya that to kill one was a capital crime. A bird so closely associated with its lofty home in the Central American cloud forest that as the forests vanished, so too did the bird. A bird that has been a symbol of liberty for several *thousand* years -- not the shrill, defiant liberty of the eagle, but the serene and innocent liberty of the child at play.

J. E. Maslow on the Resplendent Quetzal, *Bird of Life, Bird of Death*

The male is a supremely lovely bird; the most beautiful, all things considered, that I have ever seen. He owes his beauty to the intensity and arresting contrast of his coloration, the resplendent sheen and glitter of his plumage, the elegance of his ornamentation, the symmetry of his form, and the noble dignity of his carriage.

Alexander Skutch, "Life History of the Quetzal"

It was the feather that first caught my attention. An emerald green feather, long and supple, protruding from a hole in a decaying tree trunk, waving gently in the faint morning breeze. As a budding young birdwatcher, new to the American tropics and lucky enough to be wandering along a cloud forest trail in Costa Rica, I thought I was ready for anything. But not this, not such an improbable sight—a tree sprouting feathers. Suddenly that feather disappeared, swept in a flash into the hole, and a head poked out to replace it. And what a head! Topped by a gold-green, iridescent crest; fronted by a short, stout yellow beak; centered with a jet black eye, focused intently on me; and surrounding that eye, cheek to jowl, a radiant circle of shining green feathers, a jade sunburst. These were feathers on a mission, feathers like I'd never seen before. Then and there I resolved to discover what that mission was.

Left: Engraving of Resplendent Quetzal, taken from a book dated 1904

Before I could blink, the full bird emerged, launching out from its hole and rocketing off into the forest—calling a haunting, mellow *waca waca waca*—and trailing behind it a shimmering blur of emerald and ruby, its long iridescent tail unspooling like a ribbon. Now that tree-trunk feather mystery was solved: it was a tail, I could see, a very long tail, maybe a meter or more from body to tip. Clearly this tail was not an everyday phenomenon, and this was not an everyday bird. Only one bird in Central America had a tail like that, I knew. I'd just seen my first Resplendent Quetzal.

And soon I discovered it wasn't really a tail I had seen. Those long feathers streaming behind the male bird were its tail coverts, two elongated feathers that grow out above the shorter, stabilizing tail feathers (the rectrices), and cover part of their upper surface. But to most observers, myself included, it's just a "tail" that you see as the quetzal dives into the forest.

First encounters with quetzals can vary in myriad ways, a bit like first love. In my case I'd been lucky, so many elements had come together at the same time: a brilliant male Resplendent, at rest and in flight; a nest hole, glimpsed up close, with feathered life emerging from dead wood; and those haunting calls echoing on in my mind long after the bird had disappeared. Taken aback, my curiosity aroused, I knew this was a bird I had to seek out and know better. And over the years this has come to pass, with many delightful weeks in quetzal forests; many contemplative days in libraries, uncovering a trove of quetzal science and lore; and many pleasant hours in conversation with a host of people, from remote highland villages to busy urban universities, all people united in their admiration for this quetzal.

It didn't take me long to discover a few key facts about Resplendent Quetzals. They are one of five quetzal species in the genus *Pharomachrus*, all living exclusively in the Americas, confined (except for one) to highland forests, from southern Mexico to Bolivia. The Resplendent, the most northerly of the five, ranges from the mountains of Chiapas, Mexico, to those of western Panama. Like all highland quetzals, Resplendents live at tropical latitudes, though at elevations with cool temperate climates, drizzly mists, chilly and often windy nights, occasional brilliant days, and (in certain months) torrential rains. As anyone who traverses these mountains will tell you, they are all too often a land of mud. It can seem an anomaly that out of these sometimes dank and gray landscapes have sprung some of the most brilliantly plumaged birds in the world.

To understand quetzals at a deeper level, you have to start with their close relatives the trogons, a family of hole-nesting fruit eaters with compact bodies and colorful plumage. Although these tropical forest birds are also found in Africa and Asia, they achieve their highest diversity in the Americas, with 24 species identified to date. Thanks to DNA analysis, along with emerging details from geological research, it is now clear that quetzals evolved from trogons at least three to six million years ago; changing sea levels opened land bridges between North and South America, releasing trogons from their Central American confines into rich new forests, thus jump-starting a proliferation of new species. One of these likely emerged as a proto-quetzal, probably on the eastern slopes of the Andes, a bird no longer with us but many of whose genes persist in the five *Pharomachrus* quetzals that grace our world today. In this slow shaping of trogons into quetzals, we see that evolution has been nudging these birds toward ever more elaborate plumage, a more fruit-centered diet, more social behavior, and a greater reliance on nest holes in dead, rotted trees (trogons themselves make their nest holes in a variety of substrates). In many ways, quetzals can be seen as trogons that evolution has taken to extremes, culminating in the gaudy male Resplendents with their extraordinary golden crests and meter-long "tails." Central Americans speak of the Resplendent Quetzal as *el rey de los trogones*, the king of the trogons. It's hard to think of a more fitting moniker.

The four other *Pharomachrus* quetzals that reside in America's tropical forests are the Pavonine (*P. pavoninus*), a bird of hot, humid, lowland Amazonia, and the only lowland quetzal; the Golden-headed (*P. auriceps*), a cloud forest resident found along the eastern and western slopes of South America's Andes Mountains; the Crested (*P. antisianus*), which joins the Golden-headed in those same wet and windy Andean highlands; and the White-tipped Quetzal (*P. fulgidus*), a bird restricted to a narrow band of coastal mountain forest in the northern reaches of Colombia and Venezuela. All these will find a place in this book, but the Resplendent Quetzal will stay front and center, mostly because we know so much more about this bird than we do about the other *Pharomachrus*, which remain virtually unstudied.

Resplendent Quetzals have captured our imagination for millennia. Those same feathers that caught my eye in the highlands of Costa Rica in the 1980s caught other eyes going well back in history. For the complex cultures of Mexico and Central America, especially the Aztecs and Mayans a millennium and more ago, those long tail coverts of male Resplendent Quetzals took on mythical and religious status,

Left: Disappearing act. A male Resplendent Quetzal heads into its nest hole, its long tail coverts trailing behind.

A female Resplendent Quetzal gulps down one of the small avocado fruits that form a key part of its diet

A hanging bridge in the cloud forests of Monteverde, Costa Rica

their jade-green color symbolizing rebirth. Their value to these people soared, surpassing that of gold, and a robust trade sprang up to supply kings, other nobles, priests, and the military, the only sectors of society permitted to wear the garments into which these precious feathers were woven. Interestingly, it was the male covert feathers alone that had value; the bird itself was hardly known, rarely acknowledged in art and lore. And with good reason. Resplendents were confined to mountains remote from Aztec and Mayan population centers, so that, except for wandering traders, few people ever saw the birds in the wild.

But the life of this quetzal, the bird that the Aztecs barely knew, has its own allure, beyond that of the radiant feathers. As it turns out, their habits aid us in our quest to understand them. Although Resplendents can be elusive, hidden away in deep forest, they can also live quite compatibly alongside humans in more settled areas, where forests edge into field and pasture. Indeed, in such a landscape they often turn out to be surprisingly tame and easy to see, gathering in small groups at pasture edges as cows and horses graze beneath them, and even snatching fruits from backyard trees. As a result, this is a bird that people—including a small but devoted group of biologists—have gotten to know over time. A handful of studies, most undertaken in the past five decades, have begun to enlighten us about the natural history of this quetzal.

We start with the legendary ornithologist, Alexander Skutch, who spent almost a year in the early 1940s, as World War II raged in Europe, living alone in a cabin in the remote and peaceful highlands of central Costa Rica, where Resplendent Quetzals then nested abundantly. His detailed observations, focused on nesting pairs, provide the foundation for what we know about how these birds reproduce. We learn from Skutch that male and female parents share nesting duties, unusual for birds with males possessing such gaudy plumage. That the dead-tree nest holes that pairs take over from woodpeckers often need considerable retrofitting to accommodate quetzal eggs and young. That this quetzal's diet depends on just a handful of trees in the family Lauraceae, trees that produce small, fat-rich fruits (miniature avocados, essentially) that power quetzal life in their wet, chilly climate. And that, when feeding young, they supplement this with insects, lizards, snails, and other animal protein, a key addition needed to fuel the growth of nestlings—fruit alone won't do it. Skutch was also the first to suggest that some pairs

may produce two broods each year, the second quickly following the first. In short, Skutch showed us a bird with a life dependent on a limited number of trees—at most a few dozen species of live trees with oil-rich drupes that fuel Resplendent metabolism—and in each hectare of forest just a handful of nesting trees, dead and at just the right stage of rot, with holes to shelter eggs and nestlings as a new generation of quetzals takes form. It's a specialized life.

Building on Skutch's work were studies by Anne LaBastille and colleagues in Guatemala during the 1960s, research that spotlighted forest loss and other conservation issues that Resplendent Quetzals faced during those difficult years in that beleaguered country. A few decades later, efforts shifted to Monteverde, Costa Rica, and Chiapas, Mexico. A handful of their findings stand out. In Monteverde, researchers discovered some of the intricacies of how fruits and quetzals have co-evolved: how Resplendent nesting is timed to take advantage of peak fruit availability; how in years when fruit is scarce the quetzals may delay nesting, or abandon it entirely; and perhaps most importantly, how quetzals help replant the forest.

A female Collared Trogon (*Trogon collaris*), one of than 20 trogon species in the Americas. It was a trogon ancestor that gave rise to the first quetzals.

Left: Agua Azul waterfalls in Chiapas, Mexico. The Resplendent Quetzal, the most northerly of the five quetzal species, ranges from the mountains of Chiapas to those of western Panama.

Lauraceae fruits have large seeds that are too big to pass through a quetzal's digestive tract. Instead, the fruits get stripped of their outer pulp in the birds' gizzard and stomach, with the seeds then regurgitated wherever the quetzals happen to be. In just a single month, a single quetzal can cough up the seeds of hundreds of fruits. Multiply that by many quetzals over many months, and you start to see how Resplendent Quetzals (along with many other animals eating the same fruits) bring new trees to life in cloud forest habitats. These are the forests, ecological powerhouses in their own complex ways, that turn sunshine into quetzals. Quetzals return the favor, one regurgitated seed at a time, day after day and year after year.

Seasonal shifts in fruit availability are common in quetzal highlands, with the most predictable declines at the end of the breeding season, in July and August. As a result, hungry Resplendents often wander after nesting, when their young become independent. A key Monteverde study attached radio transmitters to dozens of breeding birds, tracking them throughout the year and finding that most individuals had moved off of their cloud forest nesting territories to lower elevations by July or August, finding refuge in pre-montane forests, where fruit was more available, and then wandering on to other areas before returning to their traditional nesting sites in January and February. Although the distances moved were generally short (15–40 kilometers), the elevational shifts were significant. For conservationists, these findings had important implications. No longer could nesting forests be the only focus for protection; efforts were needed in "migration" areas as well.

From Chiapas, a series of studies in the 1990s conducted by Sofía Solórzano, Mexico's premier quetzal biologist, helped confirm similar food preferences among northern Resplendents (their southern relatives in Costa Rica likely a distinct species, her data suggest), and similar tendencies to move away from breeding grounds for roughly half the year, when fruit became scarce. But perhaps most importantly her studies documented an alarming loss of cloud forest in that region. In just the three decades between 1970 and 2000, Chiapas lost almost 80% of the habitat that Resplendent Quetzals depend on, with forests cut, bulldozed, and thinned, mostly for coffee and cattle farms. Such habitat loss was not restricted to Chiapas, although reliable data have been harder to come by elsewhere. One thing was clear, however. By the early 2000s, this "king of the quetzals" was increasingly dependent on just a handful of reserves, some of them better protected than others. This is even more true today. The good news is that many of these reserves were (and remain) quite large, and quetzal populations appear to be surprisingly robust there, with breeding numbers higher than earlier

estimates had suggested. The challenge will be to keep those forests intact enough to ensure that vibrant new generations of quetzals emerge from dead trees every year.

As we are beginning to learn, cloud forests that nurture quetzals have extraordinary value, biologically as well as economically, well beyond that of the coffee and cattle that often replace them. This montane, mist-soaked habitat supports some of the highest biodiversity on the planet; in Central America alone, where just 2% of the remaining land is cloud forest, more than 5% of plant and animal species would be lost if that highland forest belt disappeared. Just as important, these forests act as a giant sponge, soaking up water from mists and rain and releasing it slowly in sparkling streams and underground aquifers, a lifeline for farms and towns at lower elevations.

Water and quetzals. Both have a role to play in encouraging cloud forest preservation. Arguably, quetzals—at least Resplendent Quetzals—could take on a more prominent role, as they are likely to spark the activism needed to head off the formidable economic pressures that threaten montane forests in Mesoamerica. I joke about Resplendents being the "million dollar bird," but in reality offering tourists the opportunity to view and photograph this bird is now a thriving business in parts of Central America, especially in that eco-tourist mecca, Costa Rica, but also in Guatemala, Panama, and Honduras. Savvy conservation groups now encourage local farmers to keep their land "quetzal-friendly," with fruit and nesting trees preserved. In return, the farmers receive a steady flow of tourist revenue for their efforts.

It is difficult to overemphasize the allure of quetzals. While you can measure some of this in tourist spending, for anyone who has glimpsed even one of these birds it is obvious there is a whole lot more involved. Resplendent Quetzals have become the "flagship species" for cloud forest conservation in Central America because of something less tangible than money. It's what they do to us. That trailing "tail"; the splendid crest; warbling calls drifting down from epiphyte-laden branches in majestic forests; the exuberance of courting males in flight on sunlit mornings; a lizard-delivering parent at a nest hole, eager young heads poking out to receive the food … there's an otherworldly aspect to this bird, something that takes us back to a time of Aztec chieftains, to a place where people and mountain wilderness were more in balance, to earth in its prelapsarian splendor. Now, the question is how do we return to that splendor?

Right: Male Resplendent Quetzals in an aerial battle, likely for mates or nest sites. Competition for both has no doubt driven the evolution of the extravagant plumage in these birds.

WHAT IS A QUETZAL?

Trogons [along with their quetzal relatives] are an intriguingly distinctive family of birds, but in truth it is the sheer beauty of their plumage that marks them out as special. In their various combinations of metallic blue or bottle-green upperparts, yellow or pink or scarlet or white underparts, meticulously patterned black-and-white undertails, and often surprisingly bright bill and eyering [sic] colors, they are as recklessly gorgeous as tropical-forest birds can be … .

Nigel Collar, in *Handbook of the Birds of the World,* vol. 6, *Mousebirds to Hornbills*

When considering the suite of closely related birds called quetzals, it is all too easy to get led astray by their mythic beauty; by their rarity; by the biologically rich and remote forests they inhabit; and by the fascinating ways their lives have intersected with humans for thousands of years. Resisting those emotions for now, we proceed more objectively in this chapter, spotlighting the essential features of these birds, introducing them one by one, and determining how they differ from one another, where they came from, and what sets them apart from other birds.

Quetzals comprise five species of mid-sized, compact-bodied, brilliantly plumaged birds in the genus *Pharomachrus,* family Trogonidae; taxonomically they exist as an offshoot of the trogons, their closest relatives. I think of them as trogons that evolution has taken to extremes, in their gaudy plumage; their diet so heavily dependent on fruit; their colonization of higher elevations with cooler climates than those that most trogons inhabit; and their tendency to gather in loose groups, particularly during courtship and breeding seasons, in areas where food is clustered and abundant.

There is one other "quetzal," a bird I choose not to consider in this book. The Eared Quetzal (*Euptilotus neoxenus*), a bird of montane pine-oak forests of northwestern Mexico and (rarely) southern Arizona and New Mexico, is the only species in its genus and until recently bore the name E*ared Trogon.* DNA evidence, egg color (same blue eggs as *Pharomachrus*), and, to the ears of some observers, vocalizations helped persuade taxonomists to move *Euptilotus* closer to the quetzals. But this bird differs substantially from *Pharomachrus*—especially in diet, behavior, and habitat—so trying to fit it into a treatment of the "true quetzals" would be awkward at best. Let us hope this bird finds a devoted Homer to sing its tale in future years.

The *Pharomachrus* quetzals are about 30% smaller than the familiar and widespread feral pigeon—and only slightly larger than North America's ubiquitous Mourning Dove (*Zenaida macroura*), known for its wistful calls. A quetzal would fit nicely in your hand and feel a little like a pigeon or dove—their bodies have the same compact shape and size, and the same soft, loose feathers. But quetzals have far more elaborate plumage than pigeons and doves, especially the brilliantly exaggerated upper wing coverts (short feathers extending from the leading edge of the wing over the tops of the flight feathers) and the elongated tail plumes, present in both sexes but far more extended in adult males, especially the Resplendents.

Female Eared Quetzal (*Euptilotis neoxenus*), a northern relative of the *Pharomachrus* quetzals.

Left: A male Resplendent Quetzal in its cloud forest habitat. The exuberance of this bird's plumage is in some ways matched by that of the vegetation around it; note the epiphyte-laden branch this bird is perched on.

The *Pharomachrus* quetzals occur from southern Mexico to Central and South America. With the exception of one species in South America, they are found in highland ("cloud") forests, where cooler, wetter conditions prevail. Thus, although they inhabit tropical latitudes (ca. 10°N to ca. 15°S), they live at elevations with temperate climates. These are lands of wind-driven mists, frequent rains, and chilly nights, interspersed with occasional brilliant sunny days, especially in the dry season. For those who live at these elevations, or visit them even briefly, there are a handful of things that improve life dramatically there, that defy the fickle weather gods: extra blankets on your bed at night, a serious rain jacket, comfortable rubber boots, and hot coffee. It seems fortuitous that the same elevations that produce quetzals also produce some of the best coffee in the world.

Before we dive into the history of these birds, their origins, and what sets them apart ecologically and behaviorally, an introduction to each of the five species will be helpful. A mere glance at these birds shows how similar they are in appearance and behavior. These five species seem to flow into each other, suggesting close relationships and recent geographic separations, that is "recent" on the evolutionary scale of hundreds of thousands of years. Such likeness is especially evident when we compare females of the five species. At a quick glance they seem almost identical; plumage differences among them are limited and subtle. Historically, taxonomists considered two of the species (Crested and Resplendent) to be one. The latter (in males at least), with their elongated tail coverts, darker eye color, and larger size can be seen as simply an exaggerated form of the Crested.

In addition, we see that despite differing geographies there are broad similarities in the habitats that support them. Step into any quetzal forest—from Mexico to Bolivia—and you are likely to notice many of the same things: a multitude of trees in fruit or flower, especially those in the family Lauraceae; epiphyte-laden branches high overhead; vines and lianas connecting ground to canopy; and a thick understory scattered with dominant emergent trees, many with buttressed roots. Those Lauraceae, with their oil-rich drupes, are in many ways the key element that has allowed quetzals to flourish and diversify. All quetzals, with the possible exception of the Pavonine, rely heavily on their fruits. As we look back at the evolution of the quetzals, we see them moving in lockstep with the Lauraceae, in all their extraordinary diversity—quetzals have followed the spread of these trees for millions of years. In Central America alone, the Lauraceae evolved more than 100 species over a few million years. As these species dispersed into emerging forests and found new ground,

quetzals were close behind. In short, without the Lauraceae, we'd probably have no quetzals today.

An astute observer in quetzal forests will note one other element, a plethora of dead, rotted trees, many riddled with woodpecker holes. Those holes, abandoned by woodpeckers and enlarged by quetzals, provide safe nest sites for quetzal eggs and young, bringing a new generation to life.

There is yet another factor linking these birds. Anyone researching quetzals soon comes to realize how little is known about them (Resplendent Quetzals excepted). Remote habitats, plus the elusiveness of these birds (they perch motionless for much of the day), have combined to discourage focused studies. For at least two of the five *Pharomachrus* quetzals, their nest and eggs have never been seen or described by scientists—a rare gap in today's ornithological knowledge.

Pavonine Quetzal
(*Pharomachrus pavoninus*)

This is a bird of hot wet lowland Amazonia, the one quetzal that has not colonized cooler, cloud-wrapped montane forests. Occupying a broad range in the upper (western) Amazon basin, millions of square kilometers overall, this bird spends its time in the middle story, the canopy, and, sometimes, at the edge of well-drained, tall, humid, *terra firme* forest (forest that is rarely, if ever, flooded). It

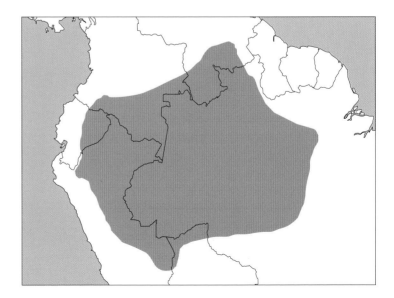

female

male

11

is occasionally found in low hills, no higher than around 700 meters in elevation. Pavonine forests extend from eastern Venezuela (southern Amazonas, Bolívar) and southeastern Colombia through the central Amazon Valley in Brazil to northeastern Peru and northern Bolivia.

Although quite similar in appearance to the Golden-headed Quetzal (*P. auriceps*), the Pavonine has a redder bill and, in males, a browner, less golden head and shorter tail coverts. It seems reasonable to think of these two quetzals as "recently" (in the last 0.5 million years or so) separated species that originated as one on the eastern slope of the Andes, with the Pavonine moving down into lowland Amazonia and establishing a separate niche there.

At 33–34 cm in length (13.4 inches) and about 160 grams (one-third of a pound) in weight, the Pavonine is roughly the same size as most other quetzals (the somewhat larger Resplendent and Golden-headed being the exceptions). Like all quetzals, adult Pavonines rely on fruit in their own diet while feeding animal protein to their nestlings to help fuel their rapid growth. At the only Pavonine nest studied (in southeastern Peru), parents brought tree frogs (Leptodactylidae) to their young about half the time they arrived at the nest; fruit made up the remaining deliveries. The Pavonine seeks out typical quetzal nest sites for its eggs: holes in dead, well-rotted trees that woodpeckers have abandoned and that the quetzals enlarge to fit their needs.

Though not recorded as abundant, Pavonines are probably not rare. But they tend to be elusive; their dense forest habitat and quiet habits make them hard to spot, so accurate surveys are a challenge. No one has reported gatherings in this quetzal, either during courtship or at fruiting trees, but here again such social behavior might have been missed. Artist-naturalist Al Gilbert, a trogon and quetzal expert whose paintings grace this book, camped for more than a month in lowland Amazonia—in Venezuela, Colombia, Brazil—and never once encountered this species, even in places with previous records.

For those who do encounter it, this is a bird more often heard than seen. Individuals usually call at dawn while moving through the forest, with vocalizations that are hard to miss and memorable (almost haunting). Steven Hilty, a dean of Neotropical bird studies and someone who has probably heard more Pavonines than any other ornithologist, describes the territorial song of this species as a slow, melancholy series of four to six long, down-slurred whistles, each preceded or followed by a sharp *chok* note, as in *wheeeear-chok*.

These vocalizations, and others described in this book, can be heard in detail at the website Xeno-canto (https://www.xeno-canto.org). Interested readers are encouraged to explore the recordings archived there.

Golden-headed Quetzal (*Pharomachrus auriceps*)

With an extensive range along both eastern and western slopes of the Andes, the Golden-headed Quetzal appears to be fairly common in forests at mid-elevations, as well as in higher-elevation cloud forests, from western Venezuela and northern Colombia south to eastern Peru and northern Bolivia. A small isolated population is located in the mountains of southeastern Panama (Cerro Pirre). Found from 1250 to 2600 meters in Podocarpus National Park, Ecuador, and from 1200 to 1500 meters in Panama, this bird overlaps broadly with the Crested Quetzal, although the latter seems to favor lower elevations. In a few cases the two species have even been known to feed in the same trees, though rarely at the same time. Such close overlap in range and (presumably) diet is unusual for species so similar in morphology and so closely related. How they

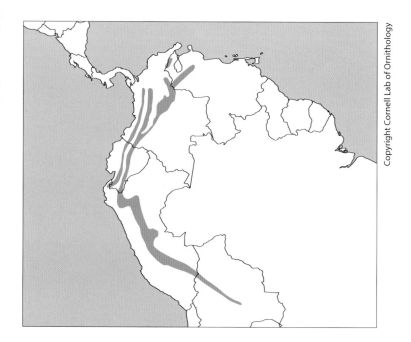

Copyright Cornell Lab of Ornithology

Right: Golden-headed Quetzal by A. E. Gilbert, courtesy of the artist.

female

male

13

maintain separate niches remains unclear and would make for an interesting dissertation.

This bird's most striking features are the burnished golden sheen of the male's head plumage, as well as the plumes (coverts) that extend just beyond the tip of the male's tail; the extended plumes hint at what the Resplendent Quetzal has taken to extremes. Interestingly, the Golden-headed (along with the Resplendent) is about 10% larger than other quetzals, perhaps because both species spend time at higher altitudes, where they encounter colder temperatures. Large size helps animals retain warmth.

In Peru, the Golden-headed is known from forests near Machu Picchu, the abandoned Inca city, leading one to wonder how Incas and this quetzal co-existed, if the bird was revered there as Resplendent Quetzals were in the Aztec and Mayan worlds, and if brilliant quetzal feathers found their way into Inca clothing and headdresses.

Not surprisingly for a quetzal, the diet of the Golden-headed skews heavily toward fruit. The stomachs of collected specimens contained 90% fruit, 10% insects. Researchers who have observed feeding birds report individuals swooping in to grab fruits in their bills, mostly larger fruits such as figs (*Ficus*), *Cecropia*, and (in the family Lauraceae) *Ocotea*, *Persea*, and *Nectrandra*, typical foods for all highland quetzals, from Mexico to Bolivia.

Joseph Forshaw, writing in his monumental book *Trogons: A Natural History of the Trogonidae* (2009), summarizes some of the key behaviors of this bird, "Usually seen singly, in pairs, or less commonly in small groups of up to 6 birds … it keeps to the mid stages and upper canopy of forest trees, where it adopts a very upright posture while sitting quietly on a horizontal branch, periodically regurgitating large seeds from recently ingested fruits … . Despite its brilliant plumage, the species is quite inconspicuous at rest and may pass undetected until it calls or makes a short, upward-swooping flight to capture a flying insect or to snatch fruits while hovering."

Along the Cordillera Central in western Colombia, the Golden-headed is thought to nest from April to June (often the rainy season there). In northeastern Ecuador, nesting apparently starts earlier, in late December and early January (generally drier). Rebecca Lohnes and Harold Greeney spent time at one nest there and provide details lacking for nearly all the other quetzals except the Resplendent. Their study found: 1) that the nest, a cavity in a dead tree about three meters above ground, was in a selectively logged forest patch, surrounded by pasture, showing that pristine forest is not always needed

to support these birds; 2) both male and female brooded the single young found in their nest, each spending roughly equal time in that effort; 3) likewise each parent brought food at roughly the same rate, with insects recorded in about half the deliveries, and fruit the rest. All this behavior is quite typical of other quetzals, the only difference being that reptiles sometimes replace insects when other quetzals are feeding nestlings.

Information on breeding also comes from captives at the Houston Zoological Gardens in Texas, US. Quetzals are notoriously difficult to breed in captivity. But the Houston researchers succeeded in getting their captive Golden-heads to lay eggs by modifying artificial nests in ways that mimicked what the birds favored in the wild: hollow logs with an appropriately sized hole, and a cavity filled with soft rotted wood that the quetzals could dig out and remove with their bills. It turns out that the digging was key; hollow logs that lacked material to be removed were ignored by the captive quetzals.

White-tipped Quetzal (*Pharomachrus fulgidus*)

With the most restricted range of any *Pharomachrus*, this quetzal is found in just a handful of mountainous regions of northern Colombia and Venezuela. White under-tail feathers distinguish this bird from others in the genus, with the male also identified by

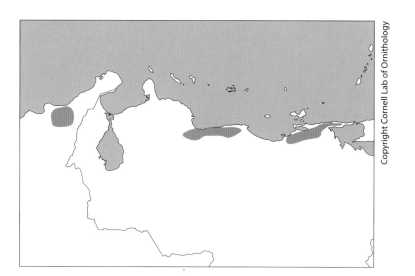

Right: White-tipped Quetzal by A. E. Gilbert, courtesy of the artist.

male

female

15

golden-bronze plumage from crown to neck (sharply set off from the green mantle [upper back] feathers), and by an inconspicuous crest of feathers at the base of the bill (known as a loral crest). In fact, this quetzal (along with the Pavonine) is the least "crested" of the quetzals—others have more prominent feather crests behind their bills, extending (in the male Resplendent) up and over the top of the head. Female White-tips have distinctive white barring on their under-tail, with the brown plumage of their breasts contrasting sharply with the red of their lower abdomen.

Although restricted in range, this is a fairly common bird within the subtropical and temperate zones it is known to inhabit; it occurs from 900 to 2500 meters, occasionally lower. A bird of humid forest and cloud forest, including clearings and edges, it is regularly seen in and around shade-coffee plantations, reminding us that growing coffee under trees has benefits for many kinds of birds, not just the migratory songbirds for which it is most promoted.

As with the other quetzals, this one is more often heard than seen, with calls described as booming hoots, and songs a rich hollow repeated *weeeeer chirú*. Before breeding, this bird can be gregarious; noisy groups of as many as 8–10 individuals have been seen feeding and displaying together. It is said to be especially vocal in the dry season (generally when quetzal breeding starts); calling birds sit upright while perched, swaying back and forth slightly as they vocalize.

Although almost nothing is known about what this bird eats, it's a good guess that, being a quetzal, fruit dominates. A few individuals have been known to feed on ripe coffee berries, and occasional lizards (likely fed to young). A recent study (2015) from northern Colombia provides a glimpse of the eggs and nest of this species: two nesting sites were described, both cavities (about 3–4 meters above ground) in dead trees and each containing two pale turquoise eggs. Unfortunately, the study did not continue long enough to describe the hatchlings, or to document their growth.

Crested Quetzal
(*Pharomachrus antisianus*)

This is another quetzal of highland Andean forests, with a narrow but extensive range almost identical to that of the Golden-heads. The Crested is found from northwestern Venezuela and Colombia south along both slopes of the Andes in Ecuador, and along the eastern

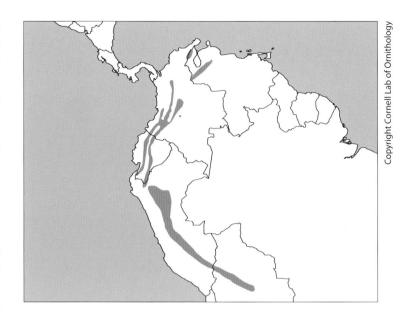

slope in Peru to west-central Bolivia. As befits its English common name, the male of this species has a prominent loral crest, a knob of green fathers protruding off the front of the head just above the bill. Females (as in all quetzals) lack such a crest. Although quite similar to female Golden-heads, female Cresteds differ in the white barring on the outer edges of their under-tail feathers and their darker eye. Female Resplendents, also quite similar in plumage, differ only in having more pronounced barring on the under-tail, and darker underparts and head.

Favoring lower elevations, the Crested often edges into thick forest, where canopy height can tower to 40 meters (130 feet), but it is also found occasionally at higher elevations in cloud and elfin forest, where trees are shorter. In northern Bolivia, where the Crested inhabits Andean foothills near La Paz and neighboring Beni, the species is fairly common at elevations between 900 and 1400 meters in tropical forest with large patches of bamboo, and in cloud forest characterized by epiphyte-laden trees, often stunted at the highest altitudes. In northern Peru, T. J. Davis found this bird to be rare at 750 meters in large tracts of undisturbed forest with canopy height of 30–35 meters; uncommon at 1050 meters in patches of ridge-top and slope forests with canopy of 20–25 meters; and missing altogether at the highest site there (1350–1450 meters). In short, the Crested appears to be generally rare wherever it is found but has a fairly broad elevational range and tolerates a variety of forest types. In northern Colombia, at 950–1050 meters, Hilty recorded it only from June to

male

female

17

August, suggesting some altitudinal movement in this species, a phenomenon, as we'll see, well known in the Resplendent Quetzal and probably occurring in other *Pharomachrus* as well.

The eggs of this species remain undescribed, and its nesting has never been studied. Few data are available on its diet; spotty observations suggest, perhaps not surprisingly, that fruits dominate, including those also taken by the Resplendent Quetzal: figs (*Ficus*) and small avocados in the genus *Ocotea*. February to June appears to be the Crested's breeding season in the Colombian Andes, a schedule that closely matches that of its resplendent relative to the north.

Resplendent Quetzal (*Pharomachrus mocinno mocinno, Pharomachrus mocinno costaricensis*)

With a range from southern Mexico to northwestern Panama, this quetzal shows interesting variation in size and plumage north to south, differences currently recognized as separate subspecies. The northern subspecies (*mocinno*) is found from Chiapas, in southern Mexico, to Nicaragua; the southern subspecies (*costaricensis*), from north-central Costa Rica to Panama. Arguably two separate species, most taxonomists continue to view *P. mocinno* as one, despite significant differences in size and genetics.

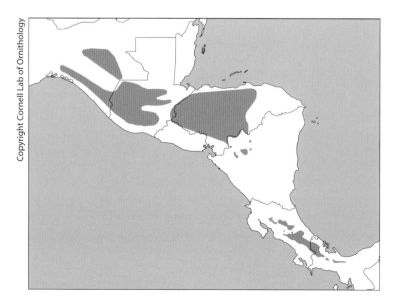

For the non-specialist, the length of male tail-plumes (coverts) has always been the most noticeable difference between northern and southern Resplendents. Those feathers are almost twice as long (and broad) in the northern birds, a startling revelation for those used to seeing this species in Costa Rica or Panama, where the male's tail feathers already seem spectacularly long. Selection, it seems, has been pushing hard on these northern males, a subject to which we will return.

For now, we note other more subtle but still significant differences between northern and southern Resplendents: body size and genetics. Sofía Solórzano, the intrepid Mexican biologist who spent much of the 1990s and early 2000s studying the species, was the first to trap, measure, and take blood from Resplendent Quetzals throughout their range (26 individuals total, all released alive). Her data showed the northern birds to be about 10% larger than their southern counterparts, both in wing length and body mass. While this may not seem like much, such consistent differences do not arise when populations are mixing. The 200+ kilometer gap between northern and southern populations—formed by a dry, hot lowland habitat in central and southern Nicaragua, a virtual desert for quetzals—is a daunting barrier for these sedentary, humid-forest birds. Solorzano's genetic data, based on blood analysis, back this up: the numbers show a genetic "distance" (% difference in genes) of at least 2%, about the same as between humans and chimpanzees. So, it's a good bet that *mocinno* and *costaricensis* have not laid eyes on each other for tens of thousand of years. It seems time for taxonomists to look more carefully at northern and southern Resplendents, with an eye toward determining their viability as separate species.

Descriptions of Resplendent Quetzals often start with the word *unmistakable*. While that is certainly true for adult males—their tail plumes alone make it difficult to mistake them for any other bird—the females are easily confused with other female quetzals. Because the Resplendent's range does not overlap that of any other quetzal, however, confusing the females is not a concern in the field. The salient features of this female are the black and white barring on the underside of the tail (only the White-tipped female has similar markings, although less crisp and distinct); the extensive gray-brown plumage of the breast and upper belly contrasting with the brilliant red of the lower belly; and the well-developed bronzy-green wing and upper-tail coverts that hint at what has reached such splendid proportions in the male.

Besides the long trailing tail plumes, males also have an extensive golden-green tufted crest of feathers from bill to top of head. Add to that the elongated and brilliant wing coverts, and the vivid

male

female

19

red plumage on the under-parts (contrasting with the white under-tail) and you have a bird that would be tough to mistake for any other.

While taxonomists may disagree on whether there are one or two species of Resplendent Quetzal, the geography of this bird is well understood. For one, its allure results in a greater number of records. In adition, it often lives in proximity to humans, on the fringes of settled areas, where forest and farmland form a mosaic and food is adequate. Except for populations in remote parks and protected areas, local people tend to know where Resplendent Quetzals can be found. Walk into any village in the highlands of Guatemala or Honduras or Panama and ask about quetzals; odds are good you will soon have a cluster of eager guides ready to escort you into the forest.

Checking range maps for this species, a few things stand out. First, the patchy nature of its distribution. Nowhere is this more evident than in northern populations (e.g., Chiapas, Mexico). Here, as in so many other parts of its range, this is essentially an "island" bird, a species living on mountaintop islands of evergreen cloud forest, most of them small. Generations of people cutting forests for cattle, coffee, and other crops have created many of these islands, or at least reduced them in size, but topography plays a role here as well. In the steep and heavily dissected mountains of Mexico and Central America, where forest type can change with just a few hundred meters shift in elevation, Resplendent Quetzals are confined to specific elevations, the ones that suit them best. Walk or drive these mountain roads and you soon become aware how quickly habitat can change. One minute you are surrounded by Resplendents in pristine cloud forest, 20 minutes later—and a change in elevation (up or down) of 400 meters or so—and you are in equally pristine forest, but the quetzals are nowhere to be found, nor the fruits they eat.

Poring over range maps shows us something else—that extensive, protected forests likely harbor the vast majority of these birds. Only where remnant forest islands give way to larger, more continuous forests, extending over dozens of square kilometers or more, does this species find the habitat it needs to survive in significant numbers. But even here the birds favor forest bands, generally between 1500 and 3000 meters, rarely higher or lower. At higher elevations, the trees tend to become stunted (so-called "elfin forests") and the colder, wetter, windier climate lacks the fruits the quetzals need. At lower elevations the desired fruits disappear or ripen on different schedules.

Four mega-forests within the range of this species stand out; combined, they probably hold at least two-thirds of all Resplendent Quetzals alive today.

Las Amistad International Park. Sitting at the southern end of the species' range, this is one of the two largest forest reserves left in Central America. The park spans southern Costa Rica and northern Panama, with 400,000 hectares (about 4000 square kilometers). It protects significant stretches of highland and lowland forest. Together with a handful of smaller protected areas along Costa Rica's Talamanca Mountains to the north (e.g., Poás Volcano, Braulio Carillo, and Chirripó national parks), this roughly 200+ kilometer corridor of montane forest likely supports 40–50% of the world's Resplendent Quetzals and at least 90% of the southern subspecies, *P. m. costaricensis*.

Bosawas Biosphere Reserve. Farther north, along the Nicaragua-Honduras border, this enormous reserve (set aside for indigenous people) protects over 2,000,000 hectares of primary forest, with perhaps 1–2% (or less) composed of highland areas that could support quetzals. We say "perhaps," because very little is known about this forest or its birds.

Sierra de las Minas Biosphere Reserve. In the highlands of southeastern Guatemala, this reserve protects about 130,000 hectares of cloud forest or 60% of this country's montane forest. By far the most important site for quetzals in Guatemala, Sierra de la Minas likely supports two-thirds of that country's Resplendent population.

El Triunfo and La Sepultura Reserves. Situated in Chiapas, Mexico, these reserves, along with some protected zones in between the two, form a relatively wild forest corridor along the western Sierra Madre, with roughly 10,000 hectares of evergreen cloud forest suitable for quetzals. Despite conservation challenges in portions of these reserves, this region helps anchor northern populations of Resplendent Quetzals.

We explore these sites and their value for quetzals in more detail in the final two chapters of this book. For now, suffice it to say that without these parks and reserves, Resplendent Quetzals would be struggling, and at significant risk overall.

NOTES FROM SOFÍA SOLÓRZANO

Researching quetzals in the field is very demanding on a personal level, since I usually work in places far from human contact. Many camp sites lack electricity, and I must often go without a hot meal, not to mention hot showers, for many days at a time. And at the end of a long exhausting day of climbing steep forest trails, I crawl into a sopping wet sleeping bag. This is work you couldn't do unless your heart is in it.

While my time in these remote cloud forests is largely a solitary existence, this isolation offers the opportunity to discover cloud forests in their most natural state. This is one of the most interesting, incredibly diverse, and rare ecosystems on the planet. I usually do research for periods of one to four months, during which time I work every day, from 6:00 am to 6:00 pm, or longer. And during this time, I will often move from one forest to another, which involves arranging for the permits and logistical support needed to make the move. As difficult as it is, one advantage of working alone is that I can move more silently as I approach a group of birds for observation. When I do change sites, I am often accompanied by an assistant and sometimes a local guide in route and then walk into the field on my own. Besides physical stamina, one must have patience in the field, for sometimes an entire day will go by without even seeing a quetzal.

When I was starting out in research, I had to learn a set of very critical skills just to be able to do my work effectively. For example, I needed to learn common quetzal vocalizations, both to help me find them and to distinguish them from the calls of other birds. I'd say it took about 3 months to learn how to identify plant food sources, and countless hours learning how to capture quetzals, place radio transmitters on them, and to draw blood from them. You could say that learning these and many other pragmatic techniques was kind of like getting a second Ph.D., one in field research methods.

I developed my own topic for my research dissertation because at that time I could find no advisors with experience in quetzals. Nonetheless, Dr. K. Oyama, who was my advisor for both my masters and doctoral work, provided great encouragement and advice during my research. And, since I initially had no funding for my research, I was very fortunate to obtain grants from the World Wildlife Fund and the American Museum of Natural History, and my university.

Of course, life being what it is, there are always challenges, though the nature of those challenges may change. In my early studies, for example, I had planned to do research for many years in Chiapas. But with the intensification of the EZLN guerilla movement in 1994, and the resulting incursion of government troops, I had to renounce that plan. I am certain now that whatever the challenges, I will always return to the field to research quetzal. That is my life's work, happily enough.

Trogons Becoming Quetzals, the Great American Interchange

To understand how quetzals became what they are today, we need to spin the evolutionary clock back to quetzal beginnings. And to do that with a modicum of accuracy, we need DNA from our five quetzal species (and from at least some of their trogon ancestors) and a geological window onto the history of the land that created them. If nothing else, this exercise is a revelation in how much information geneticists can extract from a few drops of blood, or a few tiny scrapings of skin and flesh from long-dead museum specimens. And when geologists enter the conversation, we realize to what extent the same is true for rocks, and how a small number from a few key locales can begin to help us paint a picture of the degree to which current landscapes have changed over time. Blood and rocks, the keys to unlocking millions of years of quetzal evolution.

Let's start with blood. We start in fact with trogon blood because the DNA evidence shows clearly that these are the ancestors of quetzals. Although trogons apparently originated in Eurasia some 30 to 40 million years ago, they eventually spread into Africa and east (likely via an early Bering Land Bridge) into the Americas. It was in Central America that trogons seemed to find their home 4 to 6 million years ago, rapidly evolving species in the rich and varied tropical forests taking shape at that time. Keep in mind that this was a period of particularly dynamic change on the planet, on a scale hard to imagine in our more geologically stable era. Pacific plates colliding with the American continent drove chains of volcanic eruptions, building island archipelagos that eventually coalesced into land bridges. The dramatic growth in mountains, plus shifting seas levels, made all this possible.

And it was these strips of land that would rise out of the ocean over the course of hundreds of thousands of years or so and link what we know today as North and South America. The emergent land, carpeted in new forest over millennia, became a highway along which trogons could travel, perhaps just a few dozen kilometers a year, eventually reaching much of South America. For many birds, bodies of water represent a trivial barrier. After all, hundreds of species of birds migrate each year between North and South America and between Europe and Africa, often crossing significant stretches of ocean to complete their journeys. But trogons are weak long-distance fliers. Thus the importance for trogons of forest corridors, fruit-rich paths formed during the Pliocene and Pleistocene epochs, 2 to 5 million years ago, that opened the gate to a vast forested mountain world

in South America, one that proved extraordinarily welcoming to trogons. Very likely it was one of those trogons, a species no longer with us, that evolution eventually nudged into quetzal form.

Known as the "Great American Interchange," these dispersals, each probably extending over hundreds of thousands of years, impacted far more than trogons. Mammals in particular, especially rodents, canid carnivores (the ancestors of our dogs), horses, deer, and cats, made their way south into the forests and grasslands of South America, while marsupials headed in the opposite direction, entering Central and North America from the south. With each new wave of trogons moving into South American forests, the birds radiated out into new habitat and began the slow, inevitable formation of new species. Genetic research suggests that the first proto-quetzal evolved in the Andes from these newly arriving trogons.

From that bird, new quetzal species evolved and dispersed: the White-tipped (South America's Caribbean highlands), the Pavonine (Amazon lowlands), and the Resplendent Quetzal, which radiated north, into Central America and then Mexico, reversing earlier

A male Whitehead's Trogon (*Harpactes whiteheadi*). This native of Malaysia is one of 12 species of trogon from the forests of Asia, all distant relatives of quetzals.

trogon dispersals. Many, perhaps all, of these movements appear to have tracked the spread of the lauraceous fruits that quetzals depend on. Few Neotropical fruits, if any, come as close to providing the nutrition that the oil-rich lauraceae do.

The Evolution of Beauty: The Eye of the Beholder

Supposing that quetzals are trogons that evolution has taken to extremes, then we are left with the question of how this has come to pass. What influences, over a few million years of evolution, have nudged quetzals in the directions they have taken? Why do male Resplendents have iridescent tail coverts a meter long, while trogons have survived just fine without them? Why such elaborately crested heads, why wings with brilliant and exaggerated feathers? One thing is clear: the eye-catching effects we see in all these birds are the product of what evolutionary biologists call "runaway selection," in which traits that start small become exaggerated over time.

Richard Prum, Yale professor and MacArthur Fellow, whose research focuses on birds and their dinosaur ancestors, has looked closely at the larger question here, what he calls "the evolution of beauty." Not in quetzals—he leaves them out of his discussion—but in other birds with unusually complex plumages and courtship behaviors, among them manakins (Pipridae), birds of paradise (Paradisaeidae), and peacocks (Phasianidae), in which elaborate plumage and behavior are often linked. "Our struggle to understand the origin of beauty in nature," Prum writes, "leads us back to Darwin's theory of mate choice … ," where we see "good evidence that the most refined beauty may serve as a sexual charm, and for no other purpose."

"Charm" suggests a two-way street; females respond to the incipient "beauty" that males evolve, with males then tugged further in those directions by female choice. As a result of genetic shifts and re-combinations, males randomly develop plumage and behavioral characteristics, most of which go nowhere. But a few happen to appeal to females, for reasons impossible to understand clearly. Prum explains, "Sexual selection theory holds that every elaborate ornament is the result of an equally elaborate, co-evolved capacity for aesthetic discernment … . Extreme aesthetic expression is always a consequence of extreme rates of aesthetic failure, that is, rejection by potential mates." Because some males are able to court and mate more successfully than others, the traits linked to their success—their ability to mate with a female—are the ones that carry into future generations.

Clearly the showy plumage of quetzals and other birds has a lot to do with the female eye. After all, the traits that male quetzals have developed are almost entirely visual; they have sexy plumage, for example, but not particularly elaborate vocalizations. We'll never really know what a quetzal eye takes in, and how those images are processed, but we see the results writ large in male quetzal glitter—head crests, tail and wing coverts, iridescence—sheer dazzle to catch a female's attention. And because birds are able to detect UV light, there may be even more at play here. Male quetzal feathers may show females more complexity (color, iridescence), and more individual variation, than we can perceive.

Taking a closer look at the elaborate head crests that male Resplendents have evolved, the complexity of their development leaves us in awe. Those brilliantly golden crests can be raised and lowered at will, a key part of a courting male's "branch display," the final display before mating occurs. In birds, head feathers normally grow out of skin follicles that direct the feathers toward the tail, so they lie flat on the surface of the skull and create a smooth plumage outline. But in crested birds such as Resplendent and Crested quetzals, the follicles on each side of the head have shifted orientation and point inward, so the feathers grow together and up toward the middle of the crown, meeting to create the elegant crest. The change from the smoothly feathered crown of a trogon to the regal golden crest of a Resplendent Quetzal represents thousands of tiny shifts in feather follicles, over hundreds of millennia. It is a transformation that boggles the mind.

We should note that there is no evidence of anything else at play here—no linking of crests to more robust genes for survival, to better parenting, to enhanced abilities to find food. "We now agree that ornaments evolve because individuals have the capacity, and the freedom, to choose their mates, and they choose the mates whose ornaments they prefer," Prum writes. "In the process of choosing what they like, choosers evolutionarily transform both the objects of their desires *and* the form of their own desires. It is a true evolutionary dance between beauty and desire."

Of course, it is impossible to predict where these "desires" will take a species, in what directions they will lead. The huge spreading tail of a peacock, the wire-like feathers sprouting from the head of a bird-of-paradise, and the brilliant red eye and bizarre nasal feather tuft of a Crested Quetzal—none of these could have been dreamed up by

even the most fertile human imagination (Dr. Seuss excepted?). But the female eye, linked to poorly known and complex neural networks in the brain, has nudged such "beauty" into life, choice by choice, millimeter by millimeter, millions of times over millions of years.

In some ways these transformations are best understood in the context of phylogeny, or evolutionary history, by comparing where evolution has taken "beauty" in different species of quetzals. From a single common ancestor, branched off from a trogon that no longer exists, we see the roots of different feather expressions, as well as the degree to which they have evolved. Take tail coverts. While they reach great lengths in the Resplendent Quetzal, they extend just to the end of the tail, or a few centimeters past it, in the other *Pharomachrus*. Or head crests. Here again, fully developed in the Resplendent, just partially in the Crested, and missing in the others. And yet again in the case of wing coverts, which are pronounced in the others but most exaggerated in the Resplendent. In all quetzals, then, these three feather elaborations go well beyond those of their trogon ancestors, but in the Resplendent we see much the fullest expression. Why?

To answer why quetzal plumage has moved so far beyond that of the trogons, we have to get out on a limb—or several limbs—and indulge in some speculation. So much of the behavior of these birds remains unknown. But one known difference between these two groups of birds is diet. In general, the Neotropical trogons are considerably more insectivorous than their quetzal relatives, which rely almost entirely on fruit. A fruit-centered diet means two things for quetzals. First, unlike insects, which are often camouflaged and hard to find, the fruits quetzals eat tend to be abundant and conspicuous, spread out as though a "banquet" in trees, so there's little competition among the individuals foraging there. No need to fight over feeding territories when there is plenty to go around. As a result quetzals often gather in loose groups at fruiting trees, which in the breeding season means that females have their choice of mates, and that males are more likely to compete among themselves for the chance to breed. (While trogons can also be social at times, it appears that they are less so than their quetzal relatives, with much of their lives is spent in the solitary search for insects.) In addition, because fruits tend to be relatively easy to find, quetzals have fewer demands on their time, giving rise to more elaborate courtship rituals. Hand in hand with those frequent and highly evolved courtship rituals we find elaborate plumage, as males compete for attention from choosy females and "sexy" plumage gives them an advantage.

Trying to determine why Resplendent Quetzals have evolved plumage more dazzling and complex than that of other quetzals gets us into murkier territory. We just don't know enough about the behavior of these birds, especially their courtship behavior, to craft a convincing explanation. But reading the plumage tea leaves, the odds seem good that Resplendents are more social than other *Pharomachrus*, gathering in larger groups with more intense competition among males. Their geography—that is, their northerly distribution—may have given them access to a more diverse selection of Lauraceae fruits than quetzals living farther south, giving rise to longer and more frequent gatherings. Whatever the case may be, when one watches a male Resplendent dashing into the canopy of a Guatemalan cloud forest, its meter-long tail coverts trailing behind it, it seems evident that its ecology and behavior must be at least subtly different from that of its congeners, a fascinating topic for future research.

Achieving Iridescence

As we attempt a portrait of quetzals, in all their magnificence and complexity, we find that we must spotlight one their most distinctive traits, plumage iridescence. While their trogon relatives have developed feathers with remarkably bright colors, especially reds and yellows, these colors result mostly from pigments secreted into feathers as they develop. With quetzals the dazzling colors are only partially explained by secreted pigments. Think of the glitter of a butterfly wing, or a peacock's tail, or the elytra (wing covers) of certain beetles. In these animals, their surfaces shift color in the wink on an eye, with a just a tiny change in perspective on the part of the viewer. In quetzals we find the same phenomenon.

And it is not just the Resplendent's tail. In the proper light, its emerald-gold head, throat, back, and wing coverts all glitter with iridescence. Other quetzals are not far behind. We think of the Golden-headed, with its burnished iridescent crown and upper back; the emerald loral crest and head of the Crested Quetzal; and the brilliant upper wing coverts on all the quetzals. In short, Resplendents may have run away with iridescence, but all quetzals have evolved this phenomenon to varying degrees. It's a *Pharomachrus* trait, part of the lure that males have developed to attract mates, although like many of the other aspects of quetzal "beauty" this has carried over to a lesser degree in females as well.

Left: A male Raggiana Bird-of-paradise (*Paradisaea raggiana*). This species from the forests of New Guinea epitomizes the evolution of extravagant plumage driven by male competition for mates.

The human eye is mesmerized by iridescence; science helps us understand how this glitter is brought into being. It's a fascinating and complex phenomenon involving fundamental changes in feather structure. Electron micrographs revealing cuts through portions of iridescent quetzal feathers show layers of dark-pigmented melanin platelets surrounded by tiny air spaces. Elliptical and tightly packed, the platelets are most common in the upper, dorsal portions of the feather barbule; lower down, the particles are scattered more widely. Each platelet layer is separated from the next by a layer of keratin; thus we see a melanin "sandwich" with air spaces in between.

The effect of all this is two-fold: first and perhaps most importantly, colors shift (are refracted differently) with the angle of viewing, so that in quetzals the feathers may appear golden from one angle, emerald green from a slightly different one, and even blue-violet from a third. In addition, the layers help boost the intensity of colors. Studies found that Resplendent feathers reflect certain wavelengths from each of the three layers. As Hilda Simon details in her fascinating book *Iridescence*, "the thickness of these various layers are so delicately adjusted that each reflects about the same wavelength—so that primarily one color is reinforced." Such interference demands extraordinary precision in the architecture of the feather, with variance differences among the layers of less than 1/100,000 mm! Here we glimpse the degree to which evolution has perfected this shine to attract a female eye.

There's an additional element here. When the structure of the feather is relaxed, this enhances its iridescence. In the most strongly iridescent feathers that a quetzal grows, the barbules (filaments projecting from the barb, or surface branches, of the feather) lack the tiny zipper-like hooks that hold the feather together. As a result, there is a looseness to these ornamental feathers that boosts their shimmer, as if the internal structural elements were not enough.

Despite all this, in the wet forests that quetzals inhabit their iridescence is often lost, and these birds are not beacons in the gloom. When quetzals are wet, which is often the case in cloud forests, they are surprisingly well camouflaged, blending in with the varied green foliage that surrounds them. But with a little sun, the right spot in the canopy, and maybe a flight into the open, their plumage comes alive. This should not surprise us. After all, while predators do not appear to be a major force in quetzal life (there are few records of quetzals killed by predators), they certainly exist for these birds, particularly when nesting ties them to a single locale. Continual advertising of an individual's presence would hardly benefit the bird.

Kings of the Trogons

Now that we have begun to learn about *Pharomachrus* quetzals and their descent from trogons over the last few million years, we can begin to appreciate what a rich and complex evolutionary history they have had. And we can also see that the similarities among the five species far outweigh their differences. Thanks to a succession of land bridges that spanned the Isthmus of Panama in the Pliocene epoch, releasing trogons into South America, these "kings of the trogons" began to take shape in Andean highland forests, eventually finding niches in a great variety of habitats, from cool, cloud-wrapped slopes to steamy Amazon lowlands, and even colonizing Mexican mountaintops. Thanks to a fruit-centered diet, especially fruits of the tree family Lauraceae, quetzals found opportunities to gather at abundant sources of food, social tendencies that likely promoted male competition for mates. From there, "beauty," or plumage extravagance in objective terms, took off in ways that that had never developed in their trogon ancestors. Above all, we see how one quetzal in particular, the Resplendent, was pushed by female choice to extremes, with tail plumes a meter long, a dazzling golden crest on its head, and exaggerated wing coverts that border on the hallucinogenic.

The Resplendent will be our focus for much of the rest of this book, partially due to necessity. So little is known of the other quetzals that we turn to the more available (but still limited) research on the Resplendent to carry our narrative forward. Our fate could be worse. For one thing, this bird has a fascinating life history. And, it lives in biologically rich and increasingly well-studied forests, so we can begin to appreciate its ecology in ways that are hard to do with many other birds. And, last but not least, few birds have captured the human imagination more compellingly than the Resplendent Quetzal; its life has paralleled our own for millennia.

In the next chapter we jump back a thousand years and more to the human civilizations that flourished in Middle America at that time, many of them on the edges of quetzal habitat. Resplendent Quetzals caught the eye of those people, or at least their feathers did, taking on a rich and varied role in myth and religion, especially in the Aztec and Mayan worlds.

Right: Details of feather structure and iridescence from the wing of a male Resplendent Quetzal

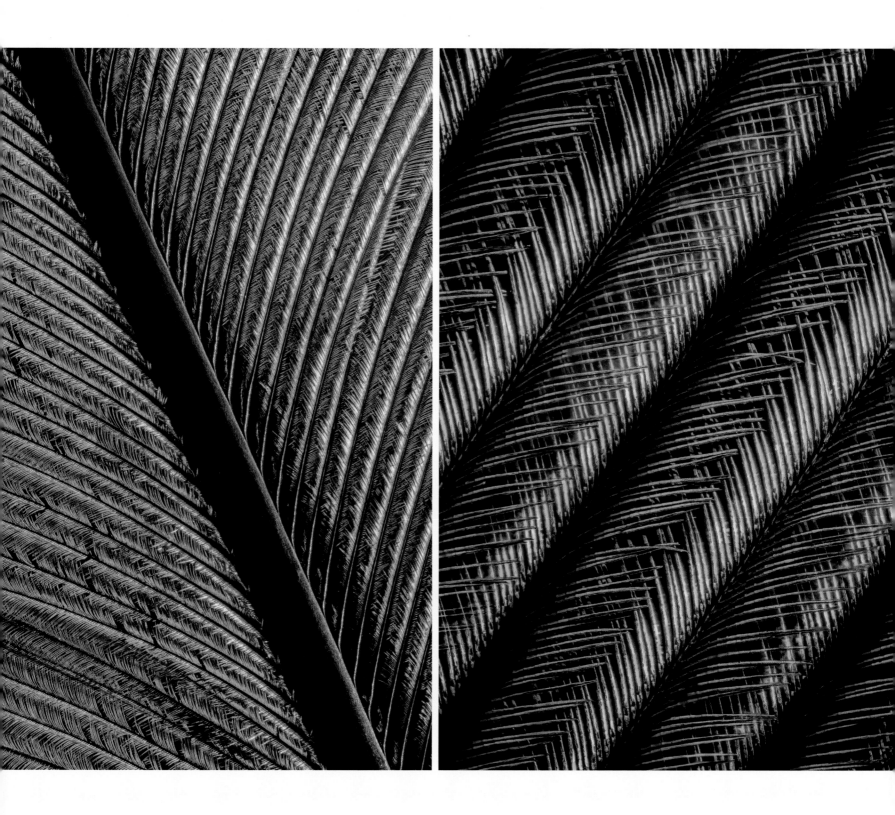

QUETZALS IN THE MAYAN AND AZTEC CIVILIZATIONS

When one embodies beings of the spiritual realm of gods and ancestors with feathers … one becomes the personification of precious life. It is as simple and as beautiful as that.

Karl Taube, *Aztec and Maya Myths,* 1993

Frances Berdan, the eminent anthropologist of Aztec civilizations, provides a fascinating account of what Europeans saw when they first arrived by ship in the Americas and waded ashore to be greeted by envoys of the royal Aztec court:

"In the spring of 1519, Hernán Cortés and his band of Spanish conquistadores feasted their eyes on the wealth of an empire. While resting on the coast of Veracruz, before venturing inland, Cortés was presented with lavish gifts from the famed Aztec emperor Moctezuma IV. While only suggestive of the vastness of imperial wealth, these presents included objects of exquisite workmanship fashioned of prized materials: gold, silver, feathers, jadeite, turquoise. There was an enormous wheel of gold, and a smaller one of silver, one said to represent the sun, the other the moon … necklaces of gold and stone mosaic work … fans and other elaborate objects created from many-colored feathers … ." The Spanish also mentioned "a piece of colored feather-work which the lords of this land are wont to put on their heads, and from it hang two ear-ornaments of stone mosaic-work with two bells and two beads of gold, and above a feather-work [piece] of wide green feathers … ."

By 1519, the Aztec imperial court was demanding tribute in the form of enormous quantities of feathers—especially the long, shimmering upper tail coverts of male Resplendent Quetzals—from the far reaches of their empire and beyond, many hundreds of kilometers from the capital city, Tenochtitlan. While it was the gold and silver that caught the greedy eyes of the conquistadors, the rare and brilliant feathers, woven into clothing unlike anything these Spanish had ever seen, were the real gifts the Aztecs laid down before their incipient conquerors. The quetzal name was part of the very fabric of their Nahuatl language. *Quetzalteuh* describes a parent's love for a child, and *quetzalchalchiuitl,* a precious type of jade stone, herb-green like the feather, adorning countless royal robes and crowns, and even placed in the mouths of the dead before burial to ensure rebirth.

From a thousand years earlier, paintings on the walls of Olmec and Mayan palaces, well to the south of the Aztec strongholds, show priests and rulers and gods with elaborately feathered headdresses, the feathers unmistakably quetzal, long and green, each with a supple curve. Clearly the appeal of the feathers—as luxury goods but also as sacred objects—spanned various cultures over at least three millennia.

Why did people find these feathers especially beautiful? How did ancient cultures meet the demand for these prized objects—and how did the quetzals survive the onslaught? For onslaught it was, with thousands of the long-plumed males hunted for feathers every year, at least at the height of the Aztec empire.

Aztecs were not the only Indigenous Americans to wear spectacular feathered headdresses. This impressive war bonnet belonged to a chieftain of exalted status in North America's Comanche tribe.

Left: Reconstructed Aztec headdress made from the elongated tail feathers (coverts) of male Resplendent Quetzals. Few such headdresses were ever made, and only the emperor and high priests were permitted to wear them, and even then only on rare occasions. These were royal feathers for a royal personage on a royal occasion.

Symbols

We will never know when the first human, or proto-human, stooped to pick up an odd bright feather on the ground and then strapped it to his or her head, feeling perhaps a little more power in their stride as they moved on to a hunt or tribal home. But we can easily imagine such a moment, and how others in the group might have quickly followed suit. Wearing feathers would have changed not only how people saw themselves, but how others saw them too. For surely this was a link that transcended the merely human, a link to the soaring power and grace of an eagle, the sunlit majesty of a courting peacock or egret, or the quick, dazzling glints of color seen on a darting cotinga or a quetzal high in the treetops, beyond the reach of earthbound mortals.

As with clothing in general, it's likely that feathers eventually helped define one's position in society, with only the wealthy—especially priests and royalty—having access to the rarest, most beautiful feathers. We think of the magnificent eagle feather headdresses worn by Native American chieftains of Great Plains tribes and of the elaborate feathered cloaks of pre-conquest Hawaiian kings, made from the feathers of thousands of small, brightly colored birds (honeycreepers). Even in our own more recent history, wealthy women's hats were, for a period in the late 19th and early 20th centuries, adorned with the plumes of egrets and other prized (but slaughtered) birds, a fashion that prompted protection, spurring the creation of the first Audubon societies and other conservation groups.

For early humans living in the land of quetzals, one feather would have been a prized find, a handful a treasure. Although the first people to find those feathers surely inhabited the highlands, where Resplendent Quetzals lived, the fascination for their feathers undoubtedly spread quickly, as more distant people discovered those precious objects through trade. And one wonders what came first, reverence for the bird or for the feather. Whatever the case, the feathers became objects of desire and worship even for those living far from the natural habitat of the quetzals. Indeed, it's likely that nearly all who wore or worshipped those feathers never saw a living quetzal.

Aside from their sheer beauty, the feathers were also rare. Nothing else like them existed in the Aztec and Mayan worlds, though jade was a close rival, with that same intense, luminous green.

But the color and size, lightness, suppleness, and iridescence made them especially coveted. The Aztec scholar Inca Clendinnen captures this magic in her writings, "[the] feather filaments are light, long, and glossy -- so that the smallest movement sets them shimmering ... And the color, a gilded emerald haunted by a deep, singing, violet blue, is extraordinary: one of those visual experiences quite impossible to bear in mind, so that each seeing is its own small miracle."

From mere beauty, it wasn't long before the feathers began to acquire symbolic value in the minds of these people, to take on an element of the sacred. The color above all, that shining iridescent green, was the color of life, of resurgent plant growth, especially corn (or maize, in Mexico and Central America), a staple grain on which much of human life depended in the Mayan and Aztec worlds.

A statue of the Mayan corn god Hun Hunahpu (known by many names), from the ancient Mayan city of Copan, about 700 CE. In many depictions of this important god, its hair was shown as green (quetzal) feathers, reminiscent of corn leaves in a sprouting field.

Left: Mural painting from the Classic Mayan site of Bonampak, in present-day Chiapas, Mexico. Quetzal covert feathers, emerald green and with their distinctive supple curve, adorn the heads of several of the warrior figures seen here, a reminder of the sacred power such feathers had for these people more than 1500 years ago.

Relief carving of the Mesoamerican god Quetzalcoatl, the feathered serpent, from a region of eastern coastal Mexico (Papantla) that paid tribute to the Aztec empire. The feathers most often associated with Quetzalcoatl were (not surprisingly given the name) those of the Resplendent Quetzal.

For those who have never visited Mexico and Central America, it might be tempting to think that the landscape there is forever green, with little seasonal change. But while the temperature may not vary much, rainfall does in many regions, with a pronounced dry season during the Northern Hemisphere's winter months, a period when many tropical trees lose their leaves and parched fields await new crops. Such drought brings a powerful thirst for rain that is difficult to explain to those who have never experienced it. In many ways it rivals the yearning for spring light and warmth that anyone who has lived through a dark cold northern winter knows all too well.

Alexander Skutch, the revered naturalist who lived for more than half a century in the Coto Brus valley of southern Costa Rica, describes this transition from dry to wet, "After a few such showers, the earth is transformed, fresh and green again, as though the stressful final days of the dry season had never oppressed it. This, not the bright, flowery days of January, is our true *primavera*, the spring when seeds germinate, resting bulbs and tubers send up new shoots, and birds sing and nest. Unlike spring at higher latitudes, the early wet season is, above all, a time of vegetative growth rather than of flowering ... ."

Little wonder that the Aztec and Mayan people responded so powerfully to this transitional time of year and saw in quetzal feathers a reflection of the vegetative renewal that was so vital to their own survival. Writing about the method of slash and burn agriculture, on which Mesoamerican peoples relied for millennia, researchers describe the sprouting of corn plants a few weeks after the first rains as forming a carpet of green "feathers" over the fields.

Corn became an essential component in the diets of Mayans and Aztecs for two key reasons. First, research has shown that the wild progenitor of corn, a grass native to the highlands of Mexico, was carefully selected by farmers in this part of the world, resulting in varieties with larger seeds, more robust yields, and a greater supply of seeds for future planting. A second development helped release key nutrients in those kernels of corn. Known today as "nixtamalization," it is a process in which lime (the mineral) is added to corn after grinding, with the mix generally soaking overnight before it is washed and then rolled out for baking or made into porridge. Ground corn prepared with lime releases a dietary source of calcium and allows the essential nutrient niacin (a form of vitamin B3) to be assimilated by the human digestive tract. Without that, and lacking other sources of niacin, humans are at risk of the debilitating and often fatal disease pellagra.

It could be argued that developing a robust and more nutritious corn crop was perhaps *the* key element allowing the Mayan and Aztec civilizations to flourish. Surplus food created a leisure class free from day-to-day tasks, giving them the time and resources to produce art, religious castes, and armies, thus creating a thriving civilization on a scale seen almost nowhere else in the Americas, and indeed in few other parts of the world, up to that time. And with such civilization came clearly defined class structures and, with those, a demand for quetzal feathers.

Right: Because the long tail (covert) feathers of male Resplendent Quetzals undulate behind them in flight, like a slithering snake, this quetzal has often been called the "snake-bird." Such a snake-quetzal association likely fostered the creation in human minds of the highly revered Mayan-Aztec god Quetzalcoatl, the feathered serpent.

In the Aztec era, ca. 1000 to 1500 BCE, trade in quetzal feathers depended on human carriers, men who walked great distances with large packs on their backs, from the mountains of present-day Guatemala to the Aztec cities of central Mexico. Here we see such carriers depicted in the Codex Mendoza, a historical collaboration (ca. 1540 BCE) between Spanish conquerors and surviving Aztec artists, which attempted to document Aztec craft traditions and their political and tribute systems. Note on several figures the use of the trumpline, a strap across the forehead that supported the weight of the cargo.

Quetzal-plumed headdresses were worn by the Olmec Maize God, the personification of young shoots of corn, with the corn silk and flowing feathers forming a natural association. A thousand years after the Olmecs, we see this same theme carried on among the Maya. On an elaborately painted plate from northern Guatemala (ca. 700–750 AD), lovely in detail, the Maize God Huun Ixi'm emerges from the earth (depicted as a turtle), wearing a headdress with curved plumes reaching halfway down his back; lesser gods beside him pour water on emerging corn shoots. Only one bird within the reach of this culture had plumes like that, and any Mayan who saw that plate would have immediately recognized the feathers.

One god emerges preeminent in these cultures, Quetzalcoatal, aka the plumed serpent. With the body of a snake and a head adorned with a sheaf of flowing quetzal feathers, this quetzal-themed deity was first depicted in early Olmec sculpture around 700–500 BCE. A thousand years later, in Mayan cities, the plumed serpent was ubiquitous in paintings and sculpture, later on adopted by a rising Aztec civilization, often with a more human form. It's a good bet that no Mayan or Aztec would have been unaware of this god or what he represented. Consider the extraordinary reach of Christ in European cultures during the Middle Ages; the quetzal-plumed serpent must have carried similar weight in Mesoamerica.

The intertwining images, in Quetzalcoatal, of snake and bird continue to busy the interpretive faculty of both archeologists and anthropologists. One might see Quetzalcoatl as a force uniting the earth (snakes being equated with the underworld) and the sky or heavens (represented by birds), and in certain cultures this may have been one element of its appeal. But, especially among the Maya, snakes were an embodiment of the heavens as much as birds were; messengers of the gods that brought rain and lightning were depicted carrying serpents in their hands.

For anyone who has seen a courting male Resplendent Quetzal in flight, its long tail plumes trailing behind it, looking snake-like as they undulate in the air, it is easy to see the connection. But depictions of flying male quetzals are missing from Mayan and Aztec art. Whatever the bird-snake connections, the universal forces behind Quetzalcoatl appear to have been centered more in the reproductive powers of the earth, in fertility and resurrection, perhaps a consolidation of the powers we see in the corn god mentioned earlier. In addition, the plumed serpent was seen as a reflection of all that was best in civilization: learning, creativity, and light—especially the dawn "star" Venus, a planet that symbolized sacred enlightenment for the Maya, and later the Aztecs.

At the Temple of Quetzalcoatl, in the Aztec city of Xochicalco in central Mexico, we see an enormous, plumed serpent carved along the side of the edifice, accompanied by an armed ruler, also thought to be associated with fertility and Venus. The latter was a patron deity of the urban center, a god of culture and civilization. So here we see Aztec rulers beginning to take on the powers and the very name of Quetzalcoatl, claiming enlightenment by stepping under the mantle of the sacred bird.

Right: This document from the Codex Mendoza shows bunches of quetzal feathers (lower left) that were brought to Aztec city markets or to the royal court. Such feather "bouquets" made for convenient articles of trade, each one having great value; many of the feather traders amassed fortunes.

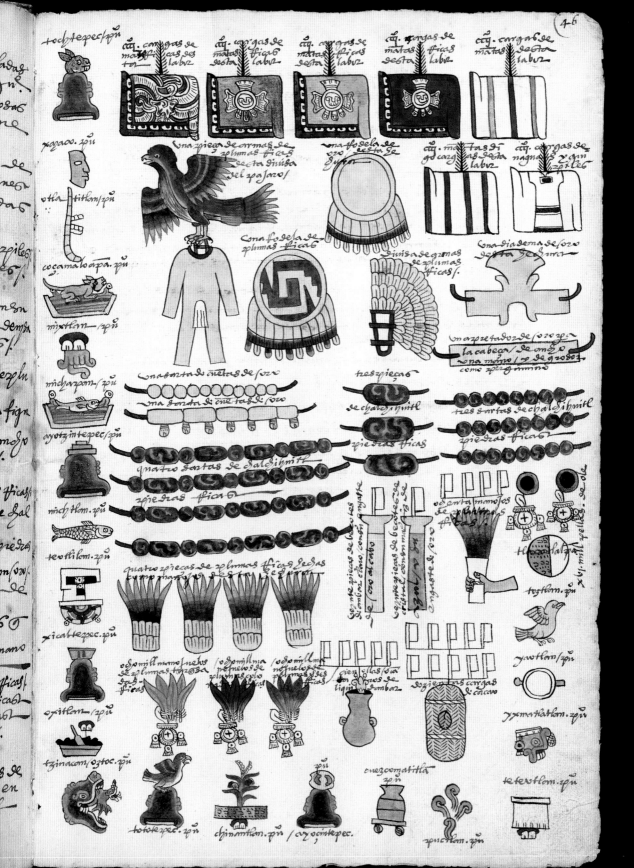

tochtepec.zpu

çcc. cargas de mantas ricas desta labor

çcc. cargas de mantas ricas desta labor

çcc. cargas de mantas ricas desta labor

çcc. cargas de mantas ricas desta labor

çcc. cargas de mantas desta labor

xayaco.zpu

otlatitlon.zpu

una pieça de armas y plumas ricas desta divisa del paxaro

una rodela de oro desta de guon

çcc. mantas de algo cargas desta labor

çcc. cargas de naguas y camiças

cocamaloapa.zpu

una rodela de plumas ricas

una rodela de oro desta de guon

una divisa de armas de plumas ricas

una diadema de oro desta de ....

mixtlan.zpu

michapan.zpu

un sartal de cuentas de oro

una sarta de cuentas de oro

tres pieças de chalchihuitl

tres sartas de chalchihuitl piedras ricas

ayotzintepec.zpu

quatro sartas de chalchihuitl

piedras ricas

piedras ricas

michtlan.zpu

teotlilan.zpu

quatro pieças de plumas ricas de las quales ...

sesenta pieças de beçotes de ... y de oro al cabo

sesenta manojos de plumas ricas

ochenta manojos de plumas ricas

xicaltepec.zpu

ochenta ... de plumas

ochenta ... nojos de plumas ... ricas

ochenta ... nojos de plumas ... ricas

cien ollas de liquidambar

dozientas cargas de cacao

yaotlan.zpu

oxitlan.zpu

tzinacanoztoc.zpu

yxmatlatlan.zpu

quecçomatitla.zpu

tototepec.zpu

chinantlan.zpu / ayoyçintepec

yputlon.zpu

Given that quetzal feathers had such a prominent role in the Aztec and Mayan worlds, one has to wonder how these people were able to get so many of them. From remote mountain cloud forests, where the birds lived and nested—but where few people lived—how were those regal Resplendent tail covert feathers (for it was those, almost exclusively, that were sought) acquired and transported to some of the most populous cities in the Americas, hundreds of kilometers away? And once in the cities, how were they transformed into garments fit for royalty?

Trade and Economics

Trying to picture the daily life of a vanished people, much less the economic fabric that held them together, all dating back some 500 years or more, is the province of the academy and Hollywood. The elaborate glyphs of Mayan script, the only Mesoamerican writing system that has been substantially deciphered, tell us little about everyday life and culture. Instead, this writing was concerned primarily with the official lives of rulers—their wars, conquests, accession to thrones, and the like—and also with astronomy and the Mayan calendar. Only for the Aztecs do we have some details of daily life, thanks to few dozen codices written after the Spanish conquest. These beautifully illustrated collaborations between Spanish and native writers and artists provide glimpses into the lives of ordinary people and what sustained them: clothing, food, religious ceremonies, medicine and herbal remedies, and tribute goods demanded by royal courts throughout the Aztec world.

In these codices, particularly the Codex Mendoza, we are given a vivid look at the prominent role that quetzal feathers played in trade during the height of the Aztec empire. In page after page of the colorful Codex Mendoza we see baskets of jade green Resplendent Quetzal feathers, or handfuls tied up in twine—all proof of the mixed lust and reverence that the Aztec court had for these prized objects. Imperial trade was based on conquest. As provinces and groups were conquered and assimilated, tribute was demanded of them, as plunder or punitively. In addition to quetzal feathers (and the feathers of other birds), a dizzying array of goods flowed into the capital, including gold, turquoise, jade, fine clothing, maize, beans, cotton, salt, lime, animal skins, live birds, timber, shells, pottery, cochineal dye, rubber, and incense.

In any one region tribute was overseen by imperial officers stationed in pueblo settlements (called *cabeceras*); the officers collected and processed goods from smaller, satellite pueblos. This state-administered system, and a multitude of informal traders, allowed urban Aztec populations to broaden their reach, exploiting widely separated regions to gather an array of valuable goods. All this trade was buoyed by the rise of an elite class, creating demand for sacred objects and exquisitely crafted articles. Such demand was ever growing, since men in elite households were allowed multiple wives (a practice forbidden to commoners). One such paterfamilias might leave behind as many as 40 wealthy households a generation or two after his death. Robust trade was thus guaranteed, and many of the traders, and the merchants to whom they sold their goods, acquired great wealth.

But transporting goods within the Aztec world was no easy thing. One had to climb mountain passes and ford lowland swamps, all without the aid of beasts of burden. While horses and oxen were integral to the lives of Europeans at this time, in pre-conquest years the people of Mexico and Central America lacked access to any mammal large enough to carry a significant load. Instead, this was a "tumpline economy." Walking on feet shod with leather sandals, steadied by robust staffs, traders and their minions followed narrow trails through dense forests, often along steep mountain ranges, carrying huge packs on their backs, supported by a strap (tumpline) around the forehead. No wheel found its way into this culture to ease that burden. Even if wheels had been available (surprisingly, Aztec children did have toys with wheels), roads would have been difficult to build in those steep and heavily forested mountainous lands.

And so there developed an entire class of merchant-traders, men who hired strong, intrepid carriers and made perilous journeys on foot to distant regions that were rich in desired goods. Many were professionals (*pochteca*) who served the Aztec emperor. They may also have doubled as spies for the royal court, with eyes and ears on the ground in provinces not yet controlled by the empire. Once outside the boundaries the emperor controlled, their lives would often have been at risk. Some were known to disguise themselves, slipping undetected into the local mix of people to trade their wares. Others apparently moved more openly and served also as ambassadors, attempting to persuade the locals of the benefits that would accrue from loyalty to the royal court, while still focused on their trade.

In one particularly well-documented example, Aztec *pochteca* traded "cotton cloaks for finely decorated clothing which they carried to Gulf coast ports of trade … returning with precious feathers (of quetzal, cotinga, red spoonbill, blue honeycreeper, yellow parrot,

Right: Artisans at work, weaving quetzal feather headdresses, likely to adorn a high priest, an emperor, or a military warrior (Florentine Codex).

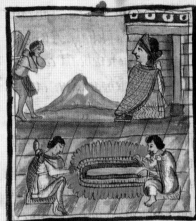

mochi tlaço ihuitl vel ipan tlapiuis.
ic nonqua quintecac, quincalten cen
tetl calli quinmacac iniscoian inana
tecahoan catca, initech pouia: nepanis
toca in tenochtitlan amanteca, ioan
in tlatilulco amanteca. Auh iniehoan
tinji, çan quiscahuiaia, inquichioaia
itlatqui. Vitzilobuchtli inquitoca
iotiaia teuquemitl, quetzalquemitl
mitzitzilquemitl, xiuhtotoquemitl,
ic tlatlacuilolli, ic tlatlatlamachilli
iniemochi iniz quican icac tlaçoih-
uitl. Yoan quichioaia iniscoian
itlatqui motecuçoma: inquin maca
ia, inquintlauhtiaia icoahoan in
altepetl ipan tlatoque, ic monotzaia
motenehoaia tecpan amanteca
itoltecahoan in tlacatl. Auh mce
quintin, motenehoaia calpiscan
amanteca, itechpouia iniz quitetl
icaca icalpiscacal motecuçoma:
iehoatl quichioaia, intlein imaceh
çallatqui motecuçoma inipan ma
cehoaia, mitotiaia: inicoac ilhuitl
quiçaia, quitlatlattitia, quitlane
nectiaia, inçaço catlehoatl queleuiz
inipan mitotiz: caceentlamantli
iecauia, cecentlamantli quichioaia

Mayan ruins at Tulum, Mexico, once an important trading hub along the Caribbean coast for quetzal feathers and other valuable goods. Many of these were carried by canoe from farther south, especially the highlands of present-day Guatemala, and off-loaded at Tulum for distribution by land to Mayan and Aztec population centers.

trogons, and unspecified green birds), jadeite (some cut), turquoise mosaic shields, many kinds of shells, tortoise-shell cups, and skins of wild animals … ." They also traded for "ornate gold accoutrements … and rock crystal ear plugs for the coastal nobility, and wares such as obsidian ear plugs, copper ear plugs, rabbit fur, and cochineal for the commoners … ."

These coastal ports along the Gulf of Mexico helped anchor a sea-going trade that flourished under the Aztecs. People on the coast mastered the art of building—and paddling—dugout canoes, many of them very large, capable of carrying a dozen paddlers and heavy loads. Such canoes had obvious advantages for traders, above all the ability to bypass the thick forest and rough swampy ground that faced anyone walking on routes inland and parallel to the coast.

For securing quetzal feathers, this coastal route became a key link between the highlands of Guatemala, where quetzals nested in abundant numbers, and the Aztec cities of central Mexico, where the feathers were worth more than gold. Feathers gathered by hunters in central Guatemalan cloud forests were sometimes carried along forest paths north and east to a large lowland lake (known today as Lago Izabal), and then on to the Caribbean by river. From there, the route along the coast of today's Belize and Yucatan Peninsula provided waters often sheltered by reefs and small islands, helping to speed the passage. Goods were eventually off-loaded for inland passage along the north coast of the Yucatan Peninsula or further north in the current Mexican states of Tabasco and Campeche.

Those coastal trading trips must have been extraordinary. Even today, in spite of growing numbers of tourists in the region, there remain wild stretches of coast with dozens of uninhabited mangrove islands (cayes), hidden sand beaches, and fish-filled lagoons. On the best of days, those ancient paddlers would have had a stiff breeze at their backs, their canoes filled with treasures, riding wind and wave through blazingly hot afternoons to camps on remote cayes along the barrier reef, making rendezvous with others plying the same route. Canoes pulled up on protected beaches at dusk; campfires with relaxed chatter around them, the scent of grilling fish rising from glowing coals; and a spray of stars across an ink-black night—all those memories would have traveled along with those paddlers as they headed out at dawn the next day, packs of quetzal feathers and other prized goods stacked carefully around their feet.

Not all the Aztec feathers traveled these coastal routes; some came along inland trails to the west, likely starting well before the Aztecs rose to power. Although reliable information is scarce, it appears that, going back as far as the Olmecs (500 BCE), traders were bringing goods to Chiapas, Mexico, in exchange for quetzal feathers, jade, and other precious items that were sourced in that region. A millennium later the Aztecs had expanded and solidified these networks.

Of course, getting the feathers was only the first step. Lovely as they were on their own, they were simply raw material waiting to be transformed into a garment fit for a priest or an emperor. Creating something of elegance, intricacy, and religious value was elaborate, time-consuming work. Perhaps not surprisingly, artisans sprang up to take this on.

In the Aztec world, each city had its guild of feather workers (known as *amanteca*). Entire districts were given over to these people, with their own schools, temples, and patron deities. These guilds had a system of internal ranking, with position and prestige based a person's ability to buy other humans—slaves—who were then sacrificed, and to conduct other religious ceremonies. Thus, wealth determined ranking; in their wealth the *amanteca* were collectively comparable to the professional merchants and traders, their caste associates.

Some feather workers wove garments and other artifacts specifically for the ruler, creating his attire and fashioning magnificent gifts for his guests. These artisans had access to the royal treasury, and thus to goods received by the emperor through tribute and foreign trade. Other artisans worked privately, probably sourcing their feathers from local traders in the open-air markets found in most large Aztec cities, and even in smaller towns. These artisans were more likely to sell to the lesser nobility, probably goods of a lower standard than those sent to the royal household.

What did they make? Sadly, feathers deteriorate quickly, as do garments decorated with them, so almost none of these treasured items survive today. But descriptions from the early days of the Spanish conquest, before the collapse of the Aztec empire, provide a glimpse of how feathers, and especially quetzal feathers, would have been transformed into what priests, rulers, and nobility might have worn every day, and when attending rituals.

Besides the extraordinary quetzal feather headdress thought to have come from Moctezuma's court (there were probably few such made on that scale), feathers were often used for ceremonial shields, handheld fans (useful for cooling and to deter biting insects), and the garments of Aztec warriors and priests. Soldiers who died in war were often buried with feathers.

Surviving the Onslaught

Although it would be an exaggeration to say that the Aztec and Mayan worlds were awash in feathers of the Resplendent Quetzal, trade in those precious items was clearly robust. But how were those feathers acquired? How were those elusive birds captured, and what

Mayan tripod vessel, ca. 400 BCE, showing a blowgun hunter shooting birds. Blowguns were the weapon of choice in the quest for Resplendent Quetzal feathers; the birds were stunned, their feathers plucked, and then released, presumably some of them unharmed.

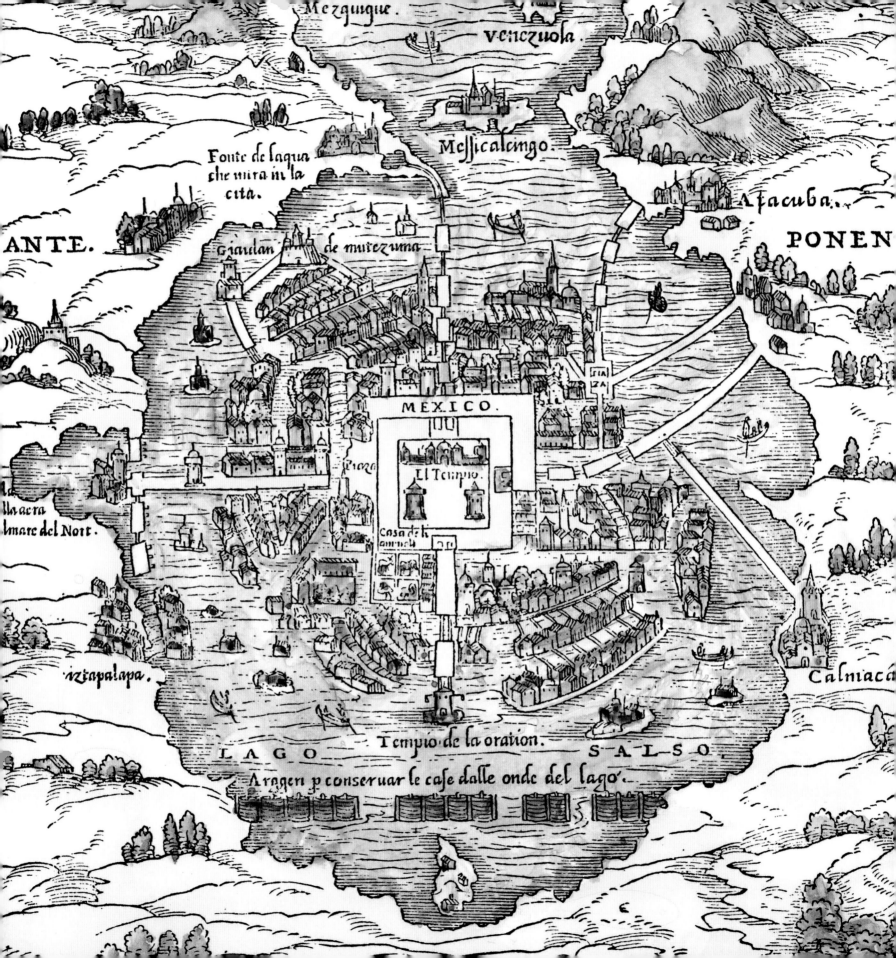

were the impacts of such "hunting" on their nesting populations? Was this a sustainable resource or did hunters deplete local populations and then move on to less disturbed, more remote ones?

Let's start with the "hunting." As trade in Resplendent Quetzal feathers accelerated, it seems likely that there were groups of men, perhaps entire villages, that specialized in finding these birds. They no doubt lived in or near prime quetzal habitat, probably on the edge of remote cloud forests at altitudes above roughly 2000 meters. Such land—often cold, wet, and cloudy—is difficult to farm, so it's a good bet that the livelihood of these people depended on finding and selling quetzal feathers. As with most such economies, they probably received just a tiny fraction of the ultimate value of the feathers.

For Mayans and the Aztecs, the Resplendent Quetzal was a sacred bird, with death the punishment for killing even a single individual. It seems likely, then, that few hunters would have knowingly killed a quetzal. Yet much of their hunting was done with blowguns, using clay pellets to stun the birds, typical of the way they hunted other birds for food. Only male quetzals were shot, their long tail coverts then plucked, and the birds (theoretically) released to grow new ones for the next breeding season.

But one wonders how easy it would be to stun a bird with a blowgun without breaking a bone, causing internal bleeding, or damaging the bird in some other way. Presumably, at least some of the quetzals hunted this way suffered long-term harm. Whatever the case, their chances of breeding successfully when robbed of their tail coverts would have been lessened.

There is some speculation that nets may also have been used to catch quetzals, probably at nest holes, where the males would have been vulnerable to such a technique. Caught in nets, the birds would have been less likely to experience harm or disruption to breeding, as nets are a gentler method than blowguns. Nevertheless, we have no clear idea how male quetzals responded to such disturbance and handling.

If sources are to be believed, a staggering number of feathers were harvested during the height of the Mayan and Aztec civilizations. Using creative detective work and a close reading of key sources, especially the Codex Mendoza, Amy and Townsend Peterson suggest that during peak years of trade the yearly harvest of feathers demanded in tribute by the royal Aztec court was affecting at least 6000 male quetzals and perhaps far more, as many as 20,000 to 30,000. After the conquest, which took place in 1575, Spanish records, likely more accurate, documented at least 15,000 to 20,000 feathers procured

annually from the Vera Paz region of central Guatemala alone. If the four long covert feathers were taken from each male, as seems likely, this means perhaps 5000 birds were caught each year in that area alone. But other areas were undoubtedly contributing feathers as well.

The effort needed to secure so many feathers annually must have been substantial. Although Resplendent Quetzals often gather in small groups near food sources, where nest sites can be especially concentrated, in many cases the birds are scattered and hard to find. In addition, they tend to live high in the forest canopy, out of reach of blowguns and nets. Even if populations then were considerably larger than they are today (and they undoubtedly were), it would have taken an "army" of hunters (in many different areas) to collect 20,000 feathers a year—many people, traveling over large swathes of cloud forest, much of it thick and tough to penetrate.

We'll never know for sure, of course, how the quetzals held up against this pressure. Initially, much of the Aztec hunting was probably focused in regions closest to the Aztec capital city of Teotihuacan, especially the Alta Verapaz of Guatemala. Current estimated populations of Resplendent populations in that region (perhaps 10,000 to 15,000 pairs) would certainly be stressed by such an Aztec take, but it's likely populations were considerably larger in the Aztec era. If so, then the capture of 3000 to 5000 males each year probably had lesser impacts on quetzal populations in that region and may even have been sustainable. But one could argue that because a significant portion of that population was nesting in montane habitats so remote that even dedicated hunters were unlikely to find them, the impact on more accessible nesting populations could have been more significant.

Overall, these hunting efforts boggle the mind. It's likely that trade networks were continually expanding to find new populations, new sources of feathers. It is not unreasonable to speculate that, at the end of this quetzal feather gold rush, as the most accessible northern populations became stressed, traders might have traveled east to Honduras and perhaps even south to Costa Rica, into the remote and lofty Talamanca Mountains, nearly 3000 kilometers from the Aztec capital.

Left: An historical map of the Aztec capital city, Tenochtitlan. As the seat of Aztec emperors, including the last true emperor, Moctezuma II, this was the destination for the vast majority of quetzal feathers harvested during the Aztec era.

QUETZAL FORESTS

The cloud forest is one place where all the capacity of language, art, and even natural science are simply not up to conveying the bewildering complexity, the infinite variety, and the daunting sense of how many of tropical nature's ways are still beyond our ken.

J. E. Maslow, *Bird of Life, Bird of Death*

A view in Costa Rica of the high peaks of the Talamanca Mountains, near the border with Panama. These mountains, supporting mostly intact forest, form one of the key remaining strongholds of the Resplendent Quetzal.

It's late March, the height of the dry season in the highlands of southern Costa Rica, and morning dawns bright and clear as I step out of a dilapidated farmhouse, my temporary home, and look up across a cow-trodden pasture to the edge of a magnificent forest. After a night of dripping clouds, with the ringing calls of frogs punctuating my sleep, the morning air is washed clean, an exhilarating clarity that can't be matched at lower elevations. Navigating rough terrain in my ill-fitting rubber boots, I weave my way slowly to the top of the pasture, avoiding a few wary cattle and passing an ancient iron waterwheel turning steadily in a rushing stream, the clank of metal and splash of water a pleasing symphony as I wake more fully into the cool mountain air. Spotting a patch of watercress, brilliant green in the crystalline water, I pluck a sprig, its bright tang a lingering taste as I clamber upslope, the sun now bringing a touch of warmth to my back.

I'm on a search for Resplendent Quetzals, but more than that I'm looking for what makes these birds thrive here. Desk-bound research has given me facts about what these birds are said to need, the key ecological elements that support a quetzal day-to-day. But without stepping into their world and staying put, settling in to take the pulse of cloud-forest life in all its complexity, I realize I will never begin to appreciate what it is about these forests that makes them such a fit for quetzals.

Left: A male Resplendent Quetzal perches in the branches of an *aguacatillo* tree, a typical resting site for this species throughout the year. These birds never stray far from their primary food source. Note the exaggerated, iridescent feathers (coverts) that extend over this male's wing, one of the plumage traits that help distinguish the Resplendent from other quetzals.

I like to think I've chosen my geography well. I stand on the edge of the Talamanca Mountains in far southern Costa Rica, prime habitat for the Resplendent Quetzal. Most of the Talamanaca highlands are protected within the borders of La Amistad International Park, a vast reserve that stretches through southern Costa Rica into western Panama. Protecting over 400,000 hectares (almost 3900 km²), this is one of the largest nature reserves in Central America and a UNESCO World Heritage Site. It's an astounding place. Recent biological expeditions into this wilderness found, in just two to three years, over 7500 species of plant, 17,000 species of beetle, and 380 species of amphibian and reptile, likely only a fraction of what's actually there in each of these categories, and the tip of the biological iceberg overall. With no roads traversing the park, and not a single hotel, hostel, or campground within it, this region is accessible to only the most intrepid of hikers. Those who know it best are the local indigenous people who have retained holdings along the park's edge, and even

Adult male Three-wattled Bellbird, whose song is a dominant sound of the Mesoamerican cloud forest during the nesting season. In this forest's avian "orchestra," it might be said that the bellbird rules the percussion section, while the Resplendent Quetzal finds a place among the woodwinds.

Stepping In

Entering this forest, stepping into cool moist air and dappled light, I'm immediately struck by the massive presence of some of the trees, a few giants that stagger the imagination. Like Dorothy in Oz, it's clear "I'm not in Kansas anymore." Shaggy trunks of gray or pale tan, lacking branches for their first 20 meters or more, rise up as immense columns from thick volcanic soils; at ground level, some of the trunks are as wide as a small pickup truck. Tipping my head back I look up at branches carpeted in green, a mini-forest of clinging epiphytic plants. Within these high gardens the bromeliads catch my eye first, many with long reddish flower spikes arcing up into forest air; their bushy pointed leaves, reminiscent of their pineapple relatives, make them easy to recognize. Orchids, many of them shriveled and parched, wait out the dry season, ready to flower with the first steady rains a few months from now. Many of the trees here seem to have more epiphytes than leaves; the epiphytes form solid green mats coating branches high over the forest floor, slowly accumulating arboreal soils.

Wind gusts catch leaves up high, and they sparkle in sunlight. I feel as if I am approaching the top of the world, and in some ways I am. If I could continue my walk to nearby peaks, a few kilometers away, and then climb a tall tree, I might be able to spot Caribbean waters to the east and then pivot around to catch a glint of sunlight off the distant Pacific to the west. Like "stout Cortez" in John Keats' poem ("On First Looking into Chapman's Homer"), I'd be tempted to imagine that I am standing "silent upon a peak in Darien." But Darien is a few hundred kilometers away, and this peak is anything but silent.

From a nearby treetop my ear is suddenly assaulted by a bizarre sound, impossibly loud—a screech precedes a metallic *klang,* sounding like a huge brass bell hit with a thick metal rod. Electronica in the cloud forest! I know it's a bird, and I know only one bird can make that call, a male Three-wattled Bellbird (*Procnias tricarunculatus*). With chestnut backs, snow-white heads and necks, and three bizarre featherless black wattles hanging from the base of their bills, these males court their drab-plumaged mates with booming *bonks* and *klangs*. No bird in the world has a louder call. The nearby male continues sounding off, his calls coming 15 to 20 seconds apart, and another male takes up

within, carving out footpaths that connect villages and remote interior farms to the outside world.

Today I'm just brushing the edge of this cloud forest wilderness; I'd never dream of making my way into it without a guide. Yet even from this limited perspective it's hard not to remain awestruck at what I see stretching up from where I stand. Geologically, these mountains are relatively young and not particularly rugged, only 2000 to 3000 meters in elevation. Fifty million years ago the highest ridges here were volcanic islands in a tropical sea, when the isthmus connecting North and South America was underwater. Since the emergence of this region, limited uplift and steady erosion from powerful rains have kept this range in check.

Even though they may not be scraping the sky, these mountains (one might better call them rugged hills) are impressive, and for me hold special relevance. From my research base in the Coto Brus region of southern Costa Rica, about 30 kilometers to the west and almost 1000 meters lower in elevation, I look up and see these peaks, steady on the horizon, often cloud-ridden, and with moisture moving in off the Caribbean on easterly winds, but on clear days seemingly close enough to touch. Whatever the weather, they always beckon with mystery—true wilderness has a way of doing that. And knowing that quetzals are there only adds to the allure.

Right: The author searching for Resplendent Quetzals in the Las Tablas Protected Zone, a prime quetzal nesting area on the edge of La Amistad International Park in southern Costa Rica.

The Black-headed Bushmaster (*Lachesis melanocephala*) is a rare but extremely venomous snake found in the lower reaches of quetzal habitat in southern Costa Rica.

the challenge, answering back; the two go back and forth, each seeming to prompt the other. When they finally quiet, I catch the ethereal, flute-like song of a Black-faced Solitaire (*Myadestes melanops*), distant thrush-music that rolls down a forested slope nearby. Like the singing of thrushes in so many parts of the world, this one stops me in my tracks, a haunting melody that seeks out the slowed-down, contemplative parts of the psyche … cathedral music for a cathedral forest.

Finally, my ear picks up the soft burry call of a quetzal and my heart beats faster. If coming from a musical instrument, it might be a low-pitched wooden whistle, melodious, with a warbling tenor. Searching out the source of this call I catch a sudden flicker of iridescent wings in a treetop, and then a long ribbon of emerald-green tail undulating behind—a male Resplendent on the move. It couldn't be anything else. Then just as suddenly the bird stops, perched quietly in the open, its tail floating ever so slightly in the faint morning breeze, its head twisting slowly to survey its surroundings. Those black beady eyes soon fix on a cluster of small, dark, red-stemmed fruits above it and, launching out from its perch, the quetzal flutters up to snatch a fruit before dropping back to its branch. Having gulped the fruit, the bird disappears into the surrounding trees.

Although it is tempting to stay put, hoping this bird will return or others will arrive to join the fruit banquet, I know of a possible nest nearby, and nests are often the best place to find these birds reliably. I trudge on up the path and soon, off to my left, spot a dead stub with a prominent hole near its top, a likely nesting spot. A nearby log, broad and moss-covered, is too inviting to ignore; it soon becomes my morning couch and makeshift blind, all in one. It doesn't take an elaborate hiding place to get close to these trusting birds. Stillness is key, and I

know my mossy resting place will let me blend into the landscape, a harmless lump on a log while the quetzals go about their lives.

Lying down and beginning to get comfortable, I look up at a high ceiling of leaves, shafts of sunlight penetrating here and there. My mind and pulse begin to slow. I've been moving too fast, I realize; for the next few hours I have nowhere to be other than where I am. The lack of human intrusion hits me first, a growing awareness that during my time in these mountain forests I've heard not a single human sound, no voices. Nor motors, not even the whine of a distant jet. Off to my side a pod of midges dances in sunlit air, a slow pulsing as they float softly up into light and then disappear down into shade, an ephemeral presence against a backdrop of massive trunks. I know this dance for what it is—a mating swarm, a gathering to ensure that fertilized females will carry this species along for another generation. Odds are good it is a species unknown to science; insects here have barely been explored. A glint of spider web catches my eye at ground level, a reminder that predators exist at many different scales.

Looking back up, I catch a flutter of pale green wings—a Tennessee Warbler (*Leiothlypis peregrina*) gleans insects from leaves high in the canopy, and I think of home and distances flown. In six weeks, I might see this same bird, hatched a summer or two ago in a Canadian forest, stopping over in my southern New England woodlot on its return to boreal nesting grounds. Writing of other birds that undertake such impressive migrations, the naturalist Peter Matthiessen reminds us not to take such journeys for granted, "One has only to consider the life force packed tight into that puff of feathers to lay the mind wide open to the great mysteries—the order of things, the why and the beginning."

A Fruitful Forest

The nest is quiet. Only one bird, a female, visits over a 50-minute period, briefly inspecting the hole and moving on. It's still early for eggs, I realize; I'll do better here in another few weeks. Reluctantly I abandon my mossy seat and begin to wander forest trails again, keeping much of my attention focused at my feet. This is, after all, a land of venomous snakes, of several species in fact, although most are rare. Still, there is no better way to ruin a day than to put your foot down on something large, lightning-quick, and extraordinarily well

Right: Bright red attachments (cupules) are conspicuous "flags" that help attract birds, including quetzals, to the fruits of many trees in the family Lauraceae.

adapted at injecting neurotoxins into the mammalian bloodstream! As I ramble on, however, snake worries fade into the more distant (but easily roused) reaches of my brain. What takes over is a growing awareness that I'm walking on fruit.

It's not a carpet of fruit, but here and there are thin scatterings, and occasionally I encounter a thicker patch. This isn't entirely new for me. At home in the New England forests I know well, autumn oak trees often drop enormous acorn crops; in some places, a hiker can crush a dozen or more acorns with every step. But here it's different. What I'm seeing is a mix of fruits, softer than acorns, a patchwork as I pick my way along the trail. Just a few fruits predominate. Some are figs, easy to recognize, coming from the most numerous and diverse trees in Neotropical forests—there could easily be 20 species of figs within just a few kilometers of where I stand. I think back to other spots I've visited in Costa Rica, where forest trails were strewn with these small oblong fruits, pale tan or green in coloration. On one occasion, softly chattering flocks of parrots fed voraciously overhead, dropping a slow rain of half-eaten figs and so "setting the table" for a host of other animals, many of them ground-sniffing, fruit-loving mammals such as agoutis, squirrels, mice, peccaries, and tapirs.

Today I note other fruits as well. Some I am unable to recognize and will have to take home to identify. Others look more familiar, especially those shaped like tiny avocados, each about the size of my smallest fingernail. They don't look like much of a meal, especially compared to the "giant" avocados on supermarket shelves back home. But these are the same fruits I saw quetzals plucking from trees a few hours earlier, and I know they play a vital role in the lives of many forest animals here, none more so than the Resplendent Quetzal. Vital because they are such prevalent trees, with perhaps 6 to 8 species in just a few hectares of forest, each on a different fruiting schedule, helping to make food available to birds through much of the year. But vital also because these fruits, *aguacatillos* (or little avocados) as they are called in Spanish, are packed with fats and nutrients; few fruits are more sustaining.

"Fruit is simply a bribe," writes the ornithologist Steven Hilty in his discussion of tropical forest birds. "Trees bribe birds with various kinds of enticing fruits in return for their services as seed dispersers. A bird eats a fruit, which has one or more seeds hidden in it, and in return for the fruit the tree's seeds are carried away unharmed to a distant location … ." "But," Hilty adds, "in the real world things are rarely that simple."

Some birds, for example, destroy seeds along with the fruits they eat. Parrots—seed predators rather than dispersers—are masters at cracking seeds open with their powerful bills and scooping out the nutritious insides with their agile tongues. Other birds crush seeds with their muscular gizzards after swallowing them. And, birds that do not destroy seeds may not carry them far, eating the fruit where they find it and then staying put, defecating or regurgitating the seeds close to where they consumed them. From the perspective of the plant, seeds dropped at its base are no more likely to survive and to grow into a mature plant than those left on the tree. Shade, competition for nutrients with the parent tree, and crowding by other seedlings sprouting around it all conspire to make it tough for such a seed to survive. Better odds go to those carried away and dropped at a

Aguacatillos, or "little avocados," are members of the laurel family (Lauraceae). They are a key component of the diet of Resplendent Quetzals throughout their range. This photograph shows *Ocotea tenera*, which, unlike most species of little avocados, is a fairly short tree (quetzals tend to browse high in the canopy).

distance, though even then germination and survival depend on a kaleidoscope of factors, including soils, light, moisture, and predators. Only a lucky few reach spots that ensure germination, and still fewer ever complete growth into a mature fruiting plant.

Nevertheless, plants increase the odds of a seed's success by evolving fruits of a color and size that attract animals most likely to be effective dispersers, such as birds that eat a lot of fruit, that don't destroy the seeds, and that tend to move about after eating. And it's not just birds. While they are key dispersers of seeds during the day, when most active, bats (especially in the Neotropics) take over at night. Like birds, they have wings, so they are agile at snatching fruit from branches and often fly considerable distances between feeding and roosting spots, where they're most likely to defecate the seeds they have swallowed along with their favorite fruits.

About 30% of Neotropical birds include fruit in their diet, a proportion roughly similar to that of birds that occur at temperate latitudes. In cooler climates, however, fruits and berries are only seasonally available, and plant diversity is relatively low, so no bird can survive through the year on fruit alone. But in the tropics, where the climate is more constant—and fruit more abundant and diverse—many birds have evolved to become fruit specialists, and depend on that food year round. For example, the members of just two Neotropical bird families—the manakins (family Pipridae) and cotingas (Cotingidae)—feed almost exclusively on more than 100 species of fruit. Quetzals and their trogon relatives are similar, straying from their fruit diet only rarely, and generally only when feeding nestlings, which require animal protein to complete their growth.

All these fruit-eating animals have a significant impact on their habitats. The combined contribution of just the birds, with perhaps hundreds in each hectare of forest depositing thousands of seeds each day, is enormous. Unwitting but efficient, they replant the forest as they go about their lives. Consider Resplendent Quetzals: 100 individuals might regurgitate as many as 75,000 seeds each month, nearly all directly in the forest. As I wandered those Talamanca highlands, I had to stop from time to time in wonder. The odds were good that many of the trees I was seeing, including some of the giants towering above me, started life riding around in the gut of a bird or a bat, or any number of ground-dwelling animals.

To remember how greatly fruits vary, we need think only of grocery store displays containing oranges, bananas, cherries, raspberries, pears, coconuts, kiwis, avocados, and so many others. But from an ecological perspective, we can separate most fruits into two general categories. There are the sweet, small-seeded fruits, the "candy" of the fruit world: raspberries, figs, tomatoes (and their wild relatives in the family Solanaceae), pokeweed (*Phytolacca* sp.), and blueberries (*Vaccinium* sp.), to name just a few. These are fruits high in carbohydrates (sugars) produced by plants living in sunny habitats such as tree-fall gaps in the forest, woodland edges and fields, the top of the forest canopy, and riverbanks. It takes sun to produce the sugars packed into these fruits, and the plants that produce them have not needed to evolve big seeds (packed with nutrients to power early growth) because their seeds tend to be deposited in sunny spots, where the sun itself powers their growth. These small seeds usually pass directly through the gut of an animal as part of the feces, a fertile matrix for getting a start on life.

Then there are the fruits with larger seeds, including avocados, cacao (*Theobroma*), tamarind (*Tamarindus*), passion fruit (*Passiflora*), and a host of other tropical species. These larger seeds, with the pulp and seed tough to separate, tend to stay in an animal's gut for a while. After the seeds are stripped of their pulp, they are often regurgitated instead of defecated; large seeds can be difficult to process through the intestines. Birds and other animals that eat these fruits are likely to live most of the time in shady forest, where such trees tend to grow. But because these bigger seeds contain considerable nutrients, packed into their endosperm, they are able to power growth even in shade, giving a seedling the start it needs to establish roots and a stem. Here again one has to marvel: evolution has perfected ways for some of the biggest trees in the cloud forest to be planted by vomiting birds.

The Lauraceae

It is fruits with larger seeds that sustain quetzals, specifically the *aguacatillos* ("little avocados" in English) I'd seen snatched from tree branches earlier in the day by fluttering Resplendents. This wasn't the first time I'd come across such foraging. Remembering the dozens of trips I'd made in search of these birds in forests like this one, more often than not the first place I found them was in a fruiting tree, especially those with canopies packed with ripe *aguacatillos*. Bird watchers seeking out Resplendent Quetzals have a mantra: find the *aguacatillos* and you'll find the quetzals. Indeed, these are the fruits that hold the key to quetzal life.

They are produced by a suite of trees in the family Lauraceae. A diverse and widespread family of plants found around the globe, the

Lauraceae grow especially well in tropical and warm temperate regions. With over 2800 species of tree, shrub, and vine, it is difficult to generalize about this family. Some are deciduous, shedding their leaves in cold or dry periods, but most are evergreen, often with aromatic leaves and bark. The family includes such familiar herbs and spices as bay laurel (*Laurus* sp.), cinnamon (*Cinnamomum* sp.), and sassafras.

When they are young, *aguacatillos* tolerate the shade produced by the trees that stand above them. But as they grow and reach the canopy and the sun, they begin to produce fruits more abundantly. Thus the fruits tend to be high above ground, favoring seed dispersers like bats and birds. Resplendent Quetzals favor the fruits produced by plants in just a few genera: *Ocotea*, *Nectandra*, *Beilschmiedia*, and *Phoebe*. These trees typically bear fruit only once each year, with 4–10 months between flowering and the development of ripe fruits, a longer period than for fruits at temperate latitudes. The flowers (generally insect-pollinated) are produced in light-colored panicles (loose, branching clusters), often hanging down or projecting upward, and often with dozens of flowers on each cluster. The oil-rich fruits (drupes) that develop tend to be numerous and easily available, often attached to their stems (pedicles) with brightly colored cupules that help to make them conspicuous. These are fruits that "want" to be seen and eaten. In the ecological literature this is referred to as "fruit-flagging," a term that refers to crops that are usually visible from a distance, luring animals to the feast. And a feast it often is, with some trees (in good years) ripening thousands of fruits over just a few weeks. Little wonder that quetzals gather in small flocks to take advantage of such bounty.

In Costa Rica alone, there are almost 130 species of trees in the Lauraceae, about half of them in the highland cloud forests that Resplendent Quetzals inhabit. Adding Mexico, Honduras, and Guatemala, this number jumps another 20 to 30%, but only a handful of species have fruits that make a real difference to quetzals. Nathaniel Wheelwright's detailed study of Resplendent diet in Monteverde, Costa Rica, found that quetzals there specialized on just 3 to 5 different lauraceous fruits throughout most of the year, all in the genera noted above. Although the Lauraceae constituted only about 4% of all bird-dispersed fruit species in the Monteverde forest, they comprised almost 50% of the fruits included in the quetzal diet there, depending on where the birds were seen feeding. At a finer scale, 80% of the fruits-seeds collected under quetzal perches (regurgitated by the birds) were lauraceous. And because these fruits were larger

than most others the quetzals ate, the proportion of lauraceous pulp in their diet was likely even higher.

Less detailed studies of Resplendent Quetzals in Chiapas, Mexico, found similar dependence on Lauraceae. By watching the birds feeding in fruiting trees, researchers there determined that a total of 15 different kinds of fruit were eaten during the peak of the nesting period (April and May), 40% of which were Lauraceae. Yet about 60% of the individual fruits consumed were in this family, which means that the birds spent more time in those trees and ate more fruit there. And just two species of Lauraceae predominated.

Resplendents consume other fruits too, particularly figs (family Moraceae), especially in the non-breeding season. In Monteverde, Costa Rica, studies showed that almost 15% of the year-round diet of these birds was figs. And farther south in highland forests, I vividly remember one of my first glimpses of these magnificent birds, some of the best looks I'd ever had of the species. At least four or five individuals were taking turns swooping out from the forest edge to grab ripe figs in the top of a tall pasture tree, plucking the fruits quickly before disappearing back into the safety of the forest, males and females alike, long fluttering tails and bright green wings parading back and forth for hours.

Resplendents also eat wild raspberries (*Morus* sp.), which can make up 5 to 10% of their diet in certain seasons and regions. Also commonly taken, at least in Costa Rica, are berries in a few key genera: *Guatteria* (Annonaceae), *Hasseltia* (Flacourtiaceae), and *Citharexylum* (Verbenaceae). But all these are quick energy fixes, hardly sustaining, perhaps the equivalent for us of grabbing a candy bar for lunch, not something that will sustain us through the afternoon, unlike the oil-rich *aguacatillos*.

If there is any doubt that fruits—especially *aguacatillos*—hold the key to Resplendent Quetzal life in the cloud forest, we need only examine this bird up close. It is a species quite literally shaped by its diet. Pick up a quetzal and right away your hand is aware of an animal packed with breast muscle, the muscle that powers flight. Dissecting specimens of these birds has shown that almost a quarter of the bird's entire body mass is flight muscle. There is good reason for this. Nearly every fruit that it eats is snatched in flight, while in a distinctive quetzal hover; and each individual does this dozens of times a day, thousands of times each year. Hanging lauraceous fruits are especially easy to harvest this way. Strong wings make it all possible, and evolution has provided the muscle to power that.

Resplendent Quetzals also have wide bills, adapted for plucking large fruits and berries, and a capacious mouth, essential for getting that fruit into the gut. Measurements show their mouth or gape is big enough to accommodate most of the *aguacatillos* that these birds are likely to find. What's more, their esophagus is thin-walled, elastic, and ringed with circular muscles, presumably helpful both in swallowing large fruits and in regurgitating seeds. From the esophagus, food enters a muscular gizzard, well adapted for grinding pulp off seeds, and then a relatively long intestine (about 50 centimeters / 20 inches), which ensures maximum extraction of nutrients from the lipid-dense pulp once it enters the digestive stream. Autopsies of Resplendent Quetzals also reveal caecal sacs (small pockets projecting off portions of the lower intestine, known in other animals to harbor digestive bacteria) packed with fruit skins, suggesting bacteria may help to process this less digestible material. In short, if we could track an *aguacatillo* through the gut of this bird, we'd see a remarkably efficient digestive process: an expandable gape and esophagus able to handle large fruits, a muscular gizzard stripping pulp from indigestible seeds, seeds remaining in the upper reaches of the gut for later regurgitation, and fruit pulp moving on through an intestine that extracts and retains nutrients.

Despite such efficiencies, there are significant nutritional limits to a diet that relies heavily on fruit. Yes, the individual fruits that Resplendents seek are often large, conspicuous, and abundant, so daily searching demands are minimal. Find the right tree or trees, settle in nearby, and energetic expenditures remain low. All quetzals lead a sedentary life. But like other fruit-specialists, these birds often face nutritional limits because much of their diet, even when the fat-rich lauraceous fruits are abundant, is barely sustaining. About 50% of a typical lauraceous fruit is indigestible seed (that avocado "pit" that we usually throw away when we cut one open) and about 75% of the surrounding pulp is water. As a result, an *aguacatillo* weighing 10 grams yields only about 1.2 grams of nutrient-dense pulp.

Even more significant, these fruits are low in protein, a key component of cellular metabolism and critical for growth and endurance. Nutritional studies of fruits eaten by Resplendent Quetzals and other birds in Monteverde, Costa Rica, found that while lauraceous fruits had a lot of lipids (fats), they contained very little protein, only 2 to 3%. Because large lauraceous fruits take up a lot of room in the gut, Resplendent Quetzals can eat only a few at a time; they must then wait to digest them, and to regurgitate the seeds, before they can eat more. Close observations of individual birds showed that visits to fruiting trees were quick, generally less than four minutes, with rarely more than 2 or 3 fruits consumed in each visit. And on average at least 30 to 40 minutes elapsed between feeding visits, during which time the birds digested pulp and regurgitated seeds. Quetzals can clearly survive on this diet, despite the chilly wet climate most inhabit, and can make it through the night, at least 12 hours of darkness, with no

food. But they have few reserves and must find food every day. Other birds, especially those that endure long migrations, have very different physiologies, storing up significant layers of fats that fuel their journeys and allow them to fast for days. Although we don't know for sure, it's a good bet that quetzals have little or no fat on their bodies.

For years, biologists thought that fruit constituted a "free lunch" for birds, especially for quetzals with their lipid-rich diet. But in fact, fruit-specialists have significant energetic limits. Resplendents, with a diet low in protein and a gut only able to accommodate a few fruits at a time, have evolved a number of adaptations to cope with these limitations. For one thing, their sedentary life keeps them from expending any more calories than necessary. More than 90% of their day is spent perched, and when they do fly, it is rarely far. In addition, when raising their young, which require protein to grow quickly, they seek out insects, frogs, and lizards to supplement the fruit that is fed to the nestlings.

We still don't know if adult Resplendents eat any such animals themselves, either when feeding nestlings or at other times. There are no data to suggest that they do, although it's hard to believe they would pass up a slow-moving beetle or a lazy lizard if either appeared within striking distance. We do know that they often delay breeding, or even skip it, in years when fruit is less abundant than usual, another adaptation to living a nutritionally challenged existence. And finally, quetzals (at least the Resplendents) adapt by "migrating" when they have to—often not far, just shifting in elevation, and almost always during the non-breeding season—all seemingly in response to shortages in local fruit availability. These movements, little studied in most populations, have generated interest for years.

On the Move

For generations, people living in quetzal highlands have known one key fact about these birds—that they tended to disappear after breeding. Some people thought the birds just went quiet and were thus harder to find, but reports in other parts of the range, generally at lower elevations, suggested the birds were finding new territory there for at least a few months each year. For Resplendent Quetzals in Central America, early reports indicated that individuals vacated breeding areas in late summer (July and August) and stayed away through

Right: Adult female Resplendent Quetzal, Costa Rica. Note the partially elongated wing and tail covert feathers, a feature that evolution has pushed to extremes in males of this species.

the autumn rainy season, returning to nesting areas sometimes as early as December but often not until February, when the trees began to flower and fruit. Some years would bring more movement than others—likely reflecting, biologists suspected, variations in the availability of fruit. But information was scarce and details few.

All this changed in the late 1980s. Pioneering studies by biologist George V. N. Powell and his collaborators in Monteverde, Costa Rica, began to unravel the mystery. In highlands on the Pacific slope, 19 breeding individuals were trapped at nests or fruiting trees and fitted with lightweight radio-transmitters, allowing the birds to go about their lives (taking fruit and feeding young) for over two years. The results were illuminating, with important implications for conservation.

This Monteverde tracking study revealed that the quetzals moved 5–20 kilometers after nesting, a relatively short distance but a significant change in elevation nonetheless. Moving just a few kilometers in that mountainous habitat can result in a change of elevation of 500 to 700 meters or more, with forest composition changing along this gradient, meaning that the birds would encounter different tree species with different fruiting schedules. By dropping down even a few hundred meters, the Monteverde quetzals were finding fruits to replace those scarce or no longer available in their nesting areas.

They first moved down the Pacific slope, leaving in July and August after their young fledged. During late summer and early fall they settled into pre-montane moist forest at around 1100–1300 meters, with scattered observations showing them feeding on lauraceous fruits. By mid-fall (October and November), they had moved upslope and east, over the Continental Divide and down into forest along the Atlantic slope at considerably lower elevations (700–1200 meters), before returning to Monteverde nesting areas in January. Although details of their diet are missing from most of this period, it seems a safe bet that these birds were seeking the same lauraceaeous fruits that they depend on the rest of the year. Once freed from parental duties after their young become independent, and with food dwindling in breeding areas, is it any surprise the Resplendents moved to places with food?

Acquiring scientific information like this—by tracking dozens of birds for months at a time—is far from easy. The radios that those Monteverde quetzals were carrying did not transmit far, perhaps a kilometer at most. And quetzal habitat in mountainous Costa Rica is tough to navigate. Planes can help, but that cloud-drenched mountainous terrain holds hazards for even the most experienced pilots. And once located from the air, biologists still had to carry receivers with bulky antennas through thick forest to get the detailed observations needed to try to determine diet and habitat. Chris Wille, a journalist who joined one of these mountainous tracking trips, provides a vivid description of the experience:

"It begins to rain as we plunge into the forest, following an old horse trail, crossing an area of steamy hot springs. The Arenal volcano is so close we can hear its belly rumbling … Vicente and Horacio move quickly; they are traveling light. Vicente has the receiver in a special waterproof case … they found a bird in this area yesterday and hope to locate it again. The trail ends and the canopy closes in; we begin climbing a steep muddy slope. The rain continues. This goes on until my knees knock with fatigue … . Vicente picks up a signal and, like a miner with a Geiger counter, dives forward. We are crossing a deep ravine. Vicente sets a killer pace, threading the antenna through the brush as easily as a buck carries it antlers. I slip and, despite Vicente's hissed warning, instinctively grab the nearest tree. It's a *pejibaye*, a type of palm with needle-like spines. My hand flashes with pain and then goes pleasantly numb, a reminder that many modern drugs, including an anesthetic used in surgery, are derived from rainforest plants … ."

These are some of the day-to-day trials and tribulations of a field researcher, a reminder of how science proceeds incrementally over months and years. But the take-home findings from this study were a revelation. Resplendent Quetzals need more habitat than anyone had anticipated. Not only did nesting forests require protection, but forests farther away were vital as well. And because these birds are reluctant to travel distances above open country, travel between nesting and "wintering" grounds must be facilitated by forest corridors. What's more, these non-breeding seasonal movements often brought the birds into areas where woodlands are more heavily fragmented than those in breeding areas, and less protected.

This was the worrisome news for conservationists. More encouraging, however, was the insight that the birds appeared to be more flexible in their choice of habitat than people had expected. For one thing the forest corridors they followed were often surprisingly narrow; it didn't take much to lead them along. And many of the quetzals adjusted well in the off season to areas were forests were more heavily fragmented, settling in to stay as long as some intact woodlands were nearby and an adequate number of fruiting trees were scattered about. Provide enough of the right food, a reliable source, and Resplendent Quetzals seem quite content to inhabit forest or pasture edge. It turns out that farming and quetzal conservation are not always mutually exclusive.

There is still a lot we don't know about the seasonal movements of Resplendent Quetzals. Follow-up studies in Monteverde could tell us if the birds still travel to the same areas they did in the 1980s and 1990s, and determine how much their movements vary year to year. Christmas bird count data from Monteverde show considerable variation in how many Resplendents remain on nesting grounds after breeding, forgoing migration and suggesting (perhaps not surprisingly) that when food is available the birds tend to stay put. All this reminds us that fruiting trees are an unpredictable resource, with significant variation in crops. Quetzals, sedentary or mobile, appear to be simply tracking fruit availability in their movements.

As brilliant as it was, the Monteverde tracking study focused on just one area and on just a single population of Resplendent Quetzals. In Chiapas, Mexico, the only other region where Resplendents have been studied year-round, researchers recorded 15 to 46 individuals in the study area during the nesting season (January to May), but only

2 to 7 from July to November, and none in August and September. Clearly, most of the birds were going elsewhere, likely to lower elevations, though details are lacking. New radios and tracking monitors make it increasingly easy to follow tagged birds. We need a handful of such studies to bolster our understanding of how Resplendent Quetzals make best use of their habitat, especially after breeding.

Simply put, Resplendent Quetzals, like most other quetzals and their trogon relatives, evolved the anatomy, behavior, and ecology required for a fruit-dominated diet because their cloud-forest habitat has abundant fat-rich fruits that meet their nutritional needs. So these quetzals came to specialize on mini-avacados, supplementing that diet only when feeding their young. Quetzals and a diverse array of lauraceous fruits evolved together through countless millennia, a co-evolutionary, mutually beneficial dance of nutrition and dispersal. These birds rely on avocado fats to keep alive in their chilly, wet montane habitats, and the forest depends on quetzals and other animals eating these fruits to disperse their seeds. Pooped and vomited seeds keep the cloud forest alive!

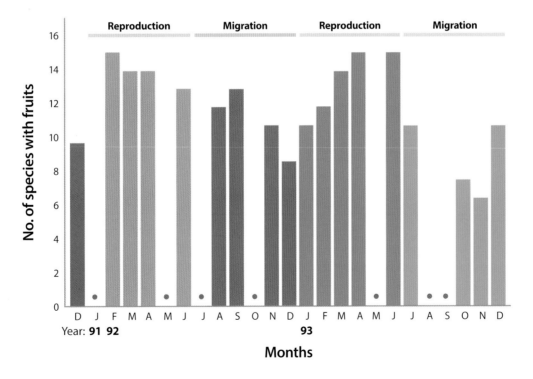

The chart shows the number of fruiting-tree species in southern Mexican (Chiapas) cloud forests, by month, in relation to the seasonal life history of Resplendent Quetzals nesting in the same forests. Note that quetzal breeding coincides with months when fruit is most available and that that they migrate when fruit is less available. (From Solórzano et al. 2000)

AT THE NEST

By mid- to late January, when the Central American dry season is in full bloom, Resplendent Quetzals become conspicuous again in their highland forest homes. Their wandering phase mostly over, they begin to follow ripening fruits upslope to habitats that will sustain their breeding over the next few months. For many individuals, these are likely familiar homes, places to which they return year after year.

As with the movements of other wild animals, schedules vary individually and year to year. Not all Resplendent populations (or individuals within populations) breed in the same week, or even month; the appearance and ripening of quetzal fruits are far from predictable. Some birds, it appears, achieve breeding condition, and find mates, more quickly than others do. But March through June forms the peak of the nesting season for nearly all Resplendent populations.

Resplendent Quetzals are the only species of quetzal for which we have more than a modicum of information on nesting and reproduction. It's likely that other *Pharomachrus* share many Resplendent breeding traits: conspicuous male courtship displays; nests in dead trees, using holes abandoned by woodpeckers; shared parental duties, with both male and female incubating eggs and feeding and brooding young; and a nestling diet that includes animal protein, a switch from the fruit-centered diet the adults depend on.

The comparative wealth of nesting information for Resplendents comes in part from a study by Alexander Skutch, who spent months watching a handful of nesting pairs in Costa Rica some 80 years ago. More recently, Nathaniel Wheelwright and others in Monteverde, Costa Rica, studied the relationship between diet and nesting, helping fill in our knowledge of what Resplendent Quetzals need to produce viable young.

It all starts with fruit. Returning Resplendents gather near fruiting trees at the start of their breeding season, and there find mates and nest sites. On a bright morning in February, if you stand on the edge of a hillside in forests that Resplendent Quetzals favor and look out over the surrounding habitat, you'll see a kaleidoscope of colors. Especially then, when new leaves are emerging and many trees are starting to flower or fruit, the forest is anything but a sea of unvarying green. From your vantage point, you'll look out over waves of color, some bright and vibrant, some pale and muted: greens and mauves, salmon and tans, dazzling yellows and pale reds. One group of trees in particular pops into view, dotted here and there across the forest, showing a blush of crimson at the top. These are the *Ocoteas*, their red-stemmed fruits dangling from canopy branches in the sun, a bright flag advertising ripeness to passing animals. These trees and others like them become the center of quetzal life as the birds settle in to find mates and nests.

Although it appears that some individuals pair up before returning to their breeding grounds (or are quick to re-establish old pair bonds), for the rest it may take longer to find a mate, sometimes a

Acorn Woodpeckers play an important role in the ecology of Resplendent Quetzals. They are prime excavators of the holes that quetzals use for nesting once the woodpeckers have abandoned them.

Left: Male Resplendent Quetzal at its nest tree

Adult male Resplendent Quetzal leaving his nest hole. Males and females both take active roles in incubating eggs and brooding and feeding young, usually sharing those tasks about equally.

month or two, and involve a swirl of activity. This is when Resplendents are most social, often gathering in loose groups, the males flying over the forest canopy in conspicuous display, or dashing through lower trees chasing females or rival males, their long tail coverts streaming behind them. Females and especially males also become more vocal. Driving forest roads in the southern highlands of Costa Rica on cool bright February mornings, I roll down the truck windows to help my aging ears pick up the distinctive warbling calls of displaying male quetzals, calls that penetrate the thick forest. Few sounds bring me greater pleasure, for I know that, with luck, my companions and I will be able to track those calls through winding woodland trails and under epiphyte-laden trees, to finally locate a quetzal gathering spot.

Left: Adult male Resplendent Quetzal (left) and female (right). Once they pair up, the two birds tend to be inseparable, feeding together and searching for (and excavating) a nest hole. By staying with his mate, the male is able to ward off other males, thus ensuring that the young he raises are his alone.

Such gatherings probably make it easier for quetzals to find mates. Once paired, male and female often perch together on a branch, from time to time making swooping, flutter-flights to snatch fruit from branches above. As with so much of quetzal life, these gatherings seem relaxed, unhurried. At least in a good year, with a bounty of fruits within easy reach, they have an easy time of it. With all that food close by and needing considerable time to digest each feeding—no wonder these birds spend so much of their lives in quiet rest.

Only with male display flights does the pace of quetzal life become more frenetic. Skutch describes one of these flights with his usual flair, "One afternoon in early March, I watched in a narrow clearing in the forest, in the midst of which stood a tall decaying trunk, where a pair of Quetzals were interested in a possible nest site. As the sun sank low, I heard mingled mellow calls and whines float out of the bordering woodland. Presently the male rushed out into the clearing, flying in a wild, dashing, irregular fashion, his long, loose, green wing-covert and tail-covert plumes vibrating madly, shouting

wac-wac-wac-wac way-ho way-ho. This appeared to be a distinct kind of flight-display, accompanied by a somewhat altered call."

Here we see those aspects of the male that literally "vibrate," catching our own attention and surely that of other Resplendents, especially females. Of course we never really know what the female quetzal eye takes in, but those wing and tail coverts are salient features in these displays, and while they are always part of the male quetzal's plumage, they come into their own, fluttering brilliance when a male takes to the air. If we think about this from an evolutionary perspective, those few minutes of display each day during the breeding period are likely a driving force in how selection has shaped this species, taking the males to extremes. Gaudy males presumably get attention, and the prize for that attention is a female willing to become a mate.

A second display may appear after the birds pair up. These so-called "branch displays" haven't been described in the scientific literature (Skutch never mentions them), but local guides who know Resplendents well can help fill in some of the details. They describe a pair perching quietly side by side when suddenly the male opens his wings, extends his prominent wing coverts outward like fingers, and leans forward over the branch, shaking his long tail plumes to complete the display. This may help cement the pair bond and stimulate copulation. Information is lacking on how regularly the birds mate afterward. We need more details to make sense of this display—and a video or two would help our understanding.

In breeding season, you may also see males chasing each other in dashing flights through the forest or above the canopy. These seem to be aggressive behaviors, triggered as competing males gather around fruiting trees. Rarely, if ever, seen outside the breeding season, they are likely demonstrations of sexual rivalry, but may have a territorial function as well.

We see, then, that fruit-laden trees bring quetzals together at the start of the breeding season. Once gathered in groups, generally loosely spaced, males attract females with aerial displays and persistent calling. Males chasing males at this time suggest that these birds are also concerned with repelling rivals and establishing nesting territories. Even after pairs form, males continue their flight displays and calling, perhaps helping to retain mates or, at failed nests, seeking new ones. This may be particularly important because some (perhaps most, we do not know) Resplendents nest twice each breeding season, the females laying a second clutch of eggs a few weeks after their first brood

Left: Young Resplendent Quetzal, about 20 days old, at their nest hole. At this stage, they eagerly await food at the nest entrance instead of hiding down in the hole, as they do when younger. Odds are good these young will be on the wing in another week.

has taken wing. Skutch was the first to broach the possibility of such double-brooding, but subsequent studies have yet to confirm it.

Finding a Nest

Once paired, male and female often search for nest sites together, and then take turns shaping a nest once they find a site they like. Wandering through quetzal forests in February or March, you might be lucky enough to spot an odd and arresting sight: clinging to the side of a decaying tree trunk, a half dozen meters or so above the forest floor, a flamboyant green bird gnawing persistently at the edges of an old woodpecker hole, with a trickle of wood chips drifting down. Like all trogons, quetzals nest in holes, and use their bills to shape them to their liking. Indeed, the trogon's name is derived from the Greek word for "gnawing."

While many trogons are capable of using their bill to dig their own nesting hole in a dead tree (or a soft wasp nest or termite mound), quetzals rely on woodpeckers to do the initial work for them, taking over holes those birds have drilled and then abandoned. Most of the woodpeckers that share highland forests with Resplendents are smaller than those gaudy birds, so it is rare for a quetzal to find a hole that does not need modification. That work can take days or weeks and requires precision; the entrance must be large enough to allow the bird to enter, but not so large that a predator can gain easy entry. Many holes initially selected prove unsuitable, the birds often abandoning one hole to try another nearby.

Both male and female Resplendents excavate nest sites. As he does so, the male clings somewhat inelegantly to a trunk, his tail blowing in the wind, taking turns with his mate to rip out bits of decaying wood with his bill. Skutch describes this nest-building activity with his usual attention to detail, "Clinging so, she bit at the decaying wood about the rim of the doorway, tearing off fairly large flakes of the soft substance and letting them drop to the ground. She continued this occupation for a minute or less, and while she was so engaged I heard soft, full notes, but could not make sure whether they arose from her or from the male perched nearby. Upon dropping away from the tree, she rejoined her waiting mate and both returned to the forest ... ."

Success for these birds depends on finding trees at just the right stage of rot, the wood soft enough to be easily ripped by a quetzal bill but not so soft that it disintegrates before an existing cavity can be hollowed out. And the tree itself must be robust enough to remain standing despite fierce highland winds. It can take weeks for quetzals to find the right one.

NEST WATCHER:
THE LEGACY
OF ALEXANDER
SKUTCH

In July of 1937, an unassuming naturalist from North America rented "an unexpectedly comfortable cottage" in the mist-filled Central Mountains of Costa Rica. A mere 33 years of age, he was already recognized for his knowledge of botany and ornithology. He had been wandering the American tropics for over a decade, studying botany and earning a living by collecting orchids and other rare tropical plants for botanical gardens and wealthy horticulturists, all the while becoming increasingly interested in tropical birds. His studies had included the nesting behavior of various trogons, but of quetzals he wrote, "I had enjoyed only fleeting glimpses on two or three occasions." Now he was eager to focus his gaze on those birds, having laid out a plan to spend 13 months doing so.

Studies by most 19th and early 20th century bird biologists were confined largely to taxonomy, with a focus on collecting specimens for museums. These were people who went into the field primarily to shoot birds, taking them back to museums and universities for study.

Skutch, by contrast, never worked for a museum, university, or government agency, and never shot or trapped a bird for science. His tools were binoculars and a notepad. He was among the first to focus on *observing* the birds of Central America, and his notes on their natural history would end up proving to be an invaluable store of knowledge.

And now Skutch would begin observing the Resplendent Quetzal and taking meticulous notes on what he saw. Quite by accident in his travels, he stumbled on an area in the highlands of Costa Rica where Resplendents were unusually abundant and easily observed. His cabin was situated on steep but relatively open pastureland abutting forest, with remnant trees in the pasture providing both the nest sites and the fruit the quetzals were looking for.

"The cottage at Vara Blanca stood on the cleared back of a narrow ridge," Skutch wrote, "with forest on either slope a short way down. From the porch I sometimes watched a pair of Quetzals foraging in the crown of a great ira rosa (*Ocotea pentagona*) that grew in the pasture on the slope to the west, its upper boughs on a level with my eyes. The birds would emerge from the forest, snatch a few of the big, green fruits in their usual dashing way, then dart down into the wooded ravine whence they had come." Complete with quetzals nesting in nearby open-pasture trees, the situation was almost too good to be true, especially considering that these birds are more typically found in deep forest.

Skutch's yearlong study of Resplendent Quetzals bequeathed to future generations a trove of data that has yet to be surpassed. Summarized in his "Life History of the Quetzal," which was published in the ornithological journal *Condor* in 1944, this remains to this day, 75 years later, the most complete account of how these quetzals go about finding mates and nest sites and taking care of their eggs and young. And his notebooks include detailed descriptions of plumages, vocalizations, nests, eggs, and the growth of young. As a fellow writer and quetzal enthusiast, I must tip my hat to this yearlong effort. It's a tour de force.

As it turned out, Skutch, who lived to be almost 100, would rarely leave the small farm he bought and settled into soon after finishing with the quetzals. There, in the Coto Brus Valley of southern Costa Rica, he was able to createt an unprecedented body of work on Costa Rican natural history, especially birds, producing more than 200 scientific papers, 20 books, dozens of articles for magazines, and four books on philosophy, all written in his lucid, charming style. Gary Stiles, author (along with Skutch) of the monumental *A Guide to the Birds of Costa Rica* (Comstock Publishing Associates, 1989), summed up how truly extraordinary this effort was:

"The legacy of Alexander Skutch to Neotropical ornithology is, quite simply, the largest body of natural-history information ever collected by a single observer. We still remain ignorant about the basic biology of most species in the world's richest avifauna, and the possibilities of obtaining that knowledge are declining—not only because of the destruction of tropical habitats, but also through our changing ways of doing and publishing science—ever less pure observation, and more deductive reasoning and statistical hypothesis testing, a trend that Alexander deplored."

Careful, patient observations were the hallmark of Skutch's ornithology. He was, like Margaret Morse Nice before him, a "watcher at the nest," brilliant at finding nests and persistent at keeping tabs on them afterward.

Skutch's writings also provide a fascinating glimpse of life on a farm in rural Costa Rica in the mid- to late 20th century. We see how he realized his goal of becoming self-sufficient, much of the food on his table coming from his own farm. We are introduced to his farmer neighbors and get a feel for how his ideas of land use and conservation often clashed with theirs. And we can't help but admire Skutch's dedicated pursuit of the simple life. This was a man who never owned a car and who connected his house to water and electricity only in the last 10 years of his life. (His wife of 50 years, devoted but long-suffering, finally prevailed upon him to buy a refrigerator at age 89.) Overarching all this was Skutch's adherence to Hindu philosophy, especially the practice of *anhisma*, doing no harm and causing no discomfort to any sentient being. He made unprecedented contributions to 20th century ornithology without once trapping or handling a bird!

Readers today can visit this farm and the forest that surrounds it, now a preserve overseen by Costa Rica's Tropical Science Center. Skutch's house and gardens remain as he left them, with trails that invite walks into nearby woods. The bird feeders are well tended, so no birder or photographer is likely to leave disappointed. Indeed, there is no better place for a naturalist to find inspiration than on the farm that Alexander Skutch left behind.

In addition to the challenges of finding and creating a nest, female Resplendent Quetzals may also need time to the gain body reserves that help produce eggs. Tending quetzal eggs and chicks takes about a month and a half, often less time than the parents spend in pairing and creating a nest, a pattern more typical for birds nesting at tropical than temperate latitudes. The latter tend to have more constrained nesting seasons, with physiologies and diets that make quicker nesting possible.

Eggs

After enlarging a nest cavity to the proper size, a female usually lays her eggs within a week or two. All this preparation, with the pair working together, may help stimulate egg-development in the female; male branch displays may help as well. In any case, as nest excavation proceeds, we see females increasingly receptive to something new, copulation. Despite the gaudy plumage, Resplendent copulations are more prosaic than one might imagine. Rarely glimpsed, a person must be lucky and persistent to see this essential sliver of quetzal life.

For me, luck arrived when I least expected it. One afternoon, while watching a female excavate a nest, apparently nearly complete, I saw her fly to a branch near where her mate was perched. The moment she landed, the male took to wing, plumes fluttering behind him, and swooped in to land on her back, his short legs struggling to maintain balance on her sleek plumage. Almost immediately, his tail scissored under hers, and within seconds it was over—the male slipping off his mate and into the air, to circle back to his perch.

Soon after a developing egg (ovum) is released into a female bird's oviduct, it must come into contact with sperm to become fertile. There is some evidence that copulation in birds may help to stimulate this release. Once fertilized, the ovum descends the oviduct during the course of one or two days, along the way picking up albumen (the egg "white"), shell membranes, and then the outer shell itself, with pigment coloring laid on just before the egg emerges into the nest. Thus the germ of new quetzal life is brought into being.

As a naturalist focused on nesting birds, I have been thinking about eggs since early in my career. As I've written elsewhere, "It is easy to take an egg for granted. They sit placidly in cartons on our refrigerator shelves, week after week, month after month. But peering into the nest of a wild bird and seeing its first few eggs, newly laid and washed with the vibrant colors of fresh pigments, is to restore a lot of the magic that supermarket eggs may have taken away from us."

Looking carefully at a quetzal egg, we can begin to appreciate all that has coaxed it into being—tiny follicles within the female, each the germ of a new egg, nurtured through the autumn rainy season by fruits gathered in sojourns to lower elevations; the return to highland forests at the start of the dry season and the stimulus of mating and nest excavation—all of these combining to help trigger a subtle flow of hormones within the female. These hormones then swell the follicles that in turn release each ovum, one by one, to become transformed within the female in a day or two into a fertile, hard-shelled ovoid sphere with a pale blue shell. Emerging into air and nestled between the warmth of the female's belly and the wood-chip lining of the nest cup, that egg, self-sufficient and alive, is poised to begin converting yolk and oxygen into a fully formed chick.

Resplendent Quetzals almost invariably lay two eggs, a day or two apart. About three times bigger and heavier than eggs of the familiar American Robin (*Turdus migratorius*) or the Central American Clay-colored Thrush (*Turdus grayi*), and a paler blue than those, they are fewer in number than the clutches of most of their trogon relatives and generally more brightly colored (trogons, like most birds that nest in dark holes, tend to lay white eggs; we don't know why hole-nesting quetzals have pigmented eggs). Each Resplendent egg weighs about 18 grams, so a clutch of two represents almost 20% of the body mass of a 200 gram female, which is not a huge demand on her resources compared with many other birds. Birds in the tropics tend to lay smaller clutches than their close relatives or ecological counterparts at higher latitudes, where the flush of spring growth often brings super-abundant food to nourish large broods.

Eggs must be tended consistently, as embryo development depends on warmth, especially hard to achieve in the quetzal's chilly mountain climate. Males and females share incubation duties during the 17 to 18 days that the eggs are developing, although the female generally spends more time on the nest. As Skutch writes, "the fundamental pattern was this: the female incubated every night and during the middle of the day; the male took a long turn [2-4 hours] on the eggs in both the morning and afternoon. Each sex was responsible for the nest twice during the cycle of twenty-four hours. But their periods of responsibility might be interrupted by brief absences, during which the eggs were left unattended. There is no reason to suppose that the female did not sleep continuously in the nest through the night; for the Quetzal, like other trogons, appears to be strictly diurnal in its activity."

Right: A male Resplendent Quetzal approaches his nest to deliver food to his young. He appears to be carrying an insect, a supplement to this bird's usual fruit diet. Young quetzals need animal protein to fuel quick growth.

Incubation rhythms—and many other behaviors—can vary among pairs. Indeed, no two Resplendent pairs are the same. In one Monteverde study by Nathaniel Wheelwright, the male attended the nest nearly three times longer than the female; in another, their duties were evenly divided. These sorts of details are revealing. We tend to think of wild animals as pretty uniform within species, each individual likely to behave very much like another. In fact, they can be as different as any two dogs within a breed, or any two human neighbors. In wild animals, these individual differences help evolution proceed, favoring the behavior of some individuals with increased survival and breeding success.

When one quetzal flies in to replace its mate, the incoming bird often gives a brief whining note as it lands at a nearby branch, or at its hole; this is usually enough to trigger the sitting bird to quit the eggs. And unlike many other trogons, little time elapses between exit and arrival. Once the cavity is cleared of the incubating partner, the bird replacing it slips in. Strangely, a departing male will often take off immediately into a flight display, calling as it goes. "Even when frightened from the nest by a passing man, the reckless bird might soar up and make himself conspicuous to all the neighborhood," Skutch writes. "Or at times he would give loud calls as he flew off, without rising above the trees." Males that continue to display, long after they have secured a mate and are tied to eggs, may still be signaling territorial claims or attracting attention from neighboring females receptive to copulation.

In Monteverde, eggs at some successful quetzal nests were left unattended for an hour or more. When food is scarce, an incubating bird may become hungry enough to leave the nest before its mate returns. But this is not always a problem. Eggs of many birds, especially those in cool climates, are remarkably resilient to chilling. Heat, especially direct sun, is far more likely to kill embryos when parents are forced to leave the nest for extended periods. Quetzal eggs may be less vulnerable to both heat and cold, sheltered in tree cavities from the worst of the elements.

Raising Young

With quetzal eggs hidden away, how can we know when they hatch? The answer is simple—the parents start bringing food to the nest. Eggs do not need to be fed, hatchlings do. Settle in near a quetzal nest, sitting quietly and keeping watch on the nest hole. Add a small dose of patience and, before you know it, you're likely to see a quetzal fly up to the nest hole with something conspicuous in its beak, perhaps a bulky beetle, a small lizard, or fruit.

What emerges from the egg of a Resplendent Quetzal? Skutch, a master at gaining access to quetzal nests, was thorough in his description of nestlings and their growth. He climbed rickety ladders, used mirrors and lights to help him peer into nest cavities, and, when possible, removed the young for short periods so he could study and photograph them.

"Like other newly-hatched trogon nestlings," he writes, "those of the Quetzal bore no vestige of down upon their pink skin. Their eyes were tightly closed. Each bore a prominent white egg tooth near the tip of the upper mandible [a hardened patch that helps the chick break out of the shell], which was slightly shorter than the lower. Their heels were studded with the short, papillate protuberances typical of nestlings that grow up in a nursery with an uncarpeted floor … When two days old, the sheaths of both their contour and flight feathers began to push through their pink skin. At four days, there was slight change, save that the nestlings were considerably larger and their feather-sheaths somewhat longer. When they were five days old, their eyelids began to separate. At eight days, they could open their eyes, but most of the time rested with the eyelids closed. On the seventh day after hatching the contour feathers of the body were breaking from the ends of their sheaths, but not those of the head. The young were ten days old before the flight plumes of the wings and tail began to push out from the tips of the sheaths, a day after the wing coverts had reached the same stage. The bill and feet were now becoming blackish."

In their first few days of life, young Resplendents are brooded by their parents almost constantly; they remain tucked into the breast feathers of the sitting bird much as the eggs were. As when incubating eggs, the parents spell each other at the nest, with the female generally taking longer shifts. Almost right away, the parents begin arriving at the nest hole with a variety of insects and other small animals, a key component of the diet of the chicks in the first week or 10 days of their lives; the parents also feed their young the lauraceous fruits that fuel adult quetzals day to day. As soon as their eggs hatch, search images shift in the brain of a parent quetzal, a shift that tells them, "find protein!"

It's easy to take the nestling phase of a bird's life for granted. We know that young birds find protection and shelter in the nest;

Left: Newly fledged Resplendent Quetzal, about 25 to 30 days old. Young at this stage can fly (poorly) but continue to be fed by the male parent. The female parent has left the family by this stage, sometimes going on to start a second brood.

that their parents brood and feed them regularly; and that they grow quickly—the nestlings of some species take flight in 10 days or less. But if we look carefully at this growth on the physiological level, where cells divide to form new tissue, the speed and complexity of this change is staggering. No one brings this phase of bird biology more alive than Julie Zickefoose, in her splendidly illustrated book *Baby Birds: An Artist Looks into the Nest.* "As much as I have witnessed it," she writes, "the transformation of a nestling each twenty-four hours is still the most astonishing yet under-appreciated natural phenomenon that I know of."

To survive, nestlings must grow rapidly—the young need to escape the nest and get on the wing, away from predators, as quickly as possible; young birds are particularly vulnerable to predators before they can fly. While many birds—mostly smaller species—grow faster than quetzals, Resplendent young still condense enormous change into a brief period, going from hatching to fledging in about 25 days. We lack precise growth measurements for these birds, but based on studies of other birds it seems likely that Resplendent chicks double their body weight in their first four days out of the egg, and then double it again in the next three or four days. Then comes their period of fastest growth, from about 7 to 12 days of age, when the nestlings are likely gaining an average of at least 4–6% of their final body weight (180–200 g) each day. Put in human terms, this would be like watching your child transforming in a just five days from a babe in arms to a child riding a bicycle.

While the young grow, their parents keep the nest clean (at least for the first two weeks), removing all droppings, which they likely

Intact cloud forest, prime habitat for Resplendent Quetzals in southern Costa Rica. Note how the treetops tend to vary in color, reflecting the fruits and flowers that different species produce. Such differences help ensure a varied diet for quetzals, over an extended period.

swallow, for they are not seen carrying away fecal sacs in their bill as do many other birds. The young at this stage tend to rest side by side on the bottom of the nest cavity, their bills pointing up. The motion of the parent as it approaches the nest hole causes the young to open their beaks and then sharply close them again, with a snap, apparently anticipating a meal. Standing near an active nest tree, you will often hear low, soft whistles—hungry young begging for food.

As the young mature, feather growth accelerates. In 14 days, a little more than halfway to full maturity and first flight, their bodies are mostly feathered, although under-wings and head still show bare skin and pinfeathers (feathers just beginning to push out of their sheaths). Flight feathers (wing and tail) are about half developed at this stage. In most birds, including quetzals, the bodies of chicks grow faster than their feathers; in the last week of the nestling period, now fully grown, most of their energy is devoted to completing feather growth, essential to getting the young birds on the wing. Compared to their trogon relatives (and possibly to other quetzal species, for which there are no comparable data), Resplendents grow their body feathers quickly, perhaps a response to the cool weather in their highland habitats.

By the time they leave the nest, the young Resplendents wear a motley plumage—blackish-brown, buff, and green—with the green, as Sketch puts it, "giving promise soon to overshadow all the others." Here we see hints of the iridescence they will begin to acquire in the next few months. Seen from a distance, these recently fledged young most closely resemble their mothers, although they lack her hints of intense green. They do have, however, the beginnings of the plumey tail coverts to come—male and female juveniles have short nubs of green above the duller tail feathers, nubs waiting to emerge as these birds mature.

Looking back at the nests from which these birds emerge there is a trace of irony in all this incipient glory. While the parents clean the nests in the first two weeks of nestling life, later, with increasing amounts of fruits being delivered, they seem to give up and allow whatever is excreted or regurgitated by the larger young to stay where it is. Researchers who have examined Resplendent nests after the young have flown, describe them as "garbage pits," reeking of feces and decaying fruit. The glory bird of Mesoamerican forests emerges from a fetid hole in a tree!

An interesting change in parental behavior occurs when the nestlings are about 14–16 days old. Up to this point, both parents have been bringing food to the nest, with the division of labor varying, sometimes considerably, from nest to nest. But after about two weeks

of feeding, the role of the female parent begins to diminish. She stops brooding the young at night (their feathers and large body size allow them to regulate their own body temperature) and brings less food. Skutch details this change, "In nearly five hours on their seventeenth day she came only thrice with food. On two of these occasions, she waited dully in the poro (*Erythrina*) tree in front of the nest, holding the morsel in her bill, until her mate arrived with food, and only then, as though stimulated by his example, did she go to the nest to deliver what she had brought. Even the preceding day, she had delayed nearly an hour, holding a green fruit until the arrival of the male caused her to take it into the nest. After this, I did not see her again in the vicinity."

It seems odd that the female parent would disappear at this stage, just a week or so before the young take to the wing. Is this not just when her offspring need food and care the most? In fact, growth of the young is slowing then, and their feeding rates too; they also no longer need brooding. The male parent is now fully capable of taking over, bringing the young enough food to push them along to their first flight at about 25 days of age.

Once the young leave the nest, they never return. They move into nearby forest, where the male parent continues to feed them, though we don't know for how long—likely not more than a week or two. Skutch witnessed this departure from the nest, "… when the young quetzals were three weeks old, I for the first time saw one of them stand on the sill of the doorway [the nest hole] and look out for a few minutes after the father had given it food … Two days later … the bird flew forth and down the slope in front of the nest … [coming] to rest about twenty-five feet about the ground. It flew well but slowly. The father … darted after the fledgling and followed it closely. For an hour, the young quetzal rested quietly on the branch where it had alighted; and here the father brought it food." Soon after this, the second young in this nest left as well, and the "family" then moved off into nearby forest where Skutch was no longer able to track them.

Why do female Resplendents abandon their nearly fledged broods, and where do they go? Lacking published data from tracked birds, we can only guess, but since some females may nest twice in a season, she may be leaving to prepare for another set of eggs—likely at a different nest than she used the first time. Caring for a first brood probably takes a toll on female Resplendents—they must lose weight hunting food for their young, and allocating more of that food to chicks than to themselves. So perhaps females leave to get back in shape for breeding, to find a mate and nest, and to develop new eggs.

We know almost nothing about the males that linger with their fledged young—how long they remain with them, how often they

feed them, and how soon the young become independent. Unlike animal prey, fruit is relatively easy to "catch," so the foraging skills that these young birds need to develop are not complicated. They need only find trees with abundant ripe fruit, which their father likely helps them locate. They are probably independent two or three weeks after first flight. Whether such a schedule gives the male parent enough time to reunite with his mate is not clear. If she re-nests, she may be more likely to find a new mate (perhaps one with less battered plumage) while her original mate is tending their first set of young. Without following marked individuals, however, answers to these questions await further study.

Nesting Highlights

While many aspects of Resplendent Quetzal nesting catch our attention, four stand out in helping to define this species. First is the conspicuous displays of breeding males. While other trogons and quetzals gather in loose assemblages and engage in chases (primarily males) through the forest, as do Resplendents, only male Resplendents (as far as we know) take off in dazzling aerial flights, an opportunity to display their conspicuous plumage to rivals nearby, and likely to current or potential mates. And they are probably the only quetzal in which males show conspicuous branch displays, as other quetzals lack comparable plumes. Second, both male and female Resplendents are persistent excavators of nest cavities in decaying wood, usually taking advantage of holes that woodpeckers have created and abandoned. While their trogon relatives also nest in holes, and are competent excavators in a variety of substrates, none seem to specialize in the very soft wood that Resplendents and other quetzals need—dead trees in their final stages of decay. Third, male Resplendents are reliable mates, active participants in the care of eggs and young, especially as the young approach fledging. At that stage, they take over feeding the fledglings, presumably leading them to suitable habitat—and allowing their mates to move on to a second brood, if that occurs. Lastly, in no other quetzal or trogon has it been shown how closely the breeding season coincides with peak fruiting in surrounding forests. Future studies may demonstrate that others are similarly synchronous, but with Resplendents it is clear why they nest during dry season months—fruit is most available then.

Left: Heads and tails. Because male Resplendents have such long tail coverts, their plume-tips often projects out from the nest.

THE ONCE AND FUTURE QUETZAL

For anyone inquisitive, initial fascination with quetzal life inevitably gives way to concern about what lies in store for these magnificent birds. Given what we know today, can we predict with any degree of certainty the future of these five species, a decade or two from now, not to mention a century hence? There are no easy answers here. Not surprisingly, the farther out we look, the less we can say. Although there are good reasons to be optimistic in the near term, there are worrisome clouds on the horizon.

I remember my first glimmer of optimism about where quetzals might be headed. I was standing on a hillside in the highlands of central Costa Rica, looking up at a tree full of the small wild avocadoes that Resplendent Quetzals are known to rely on. Surrounded by a pleasing mosaic of forests and small farms, I was on the edge of a tiny field, steep but carefully terraced, clearly well-tended. The cabbages growing there were magnificent, plump, and shiny in the morning sun, and the farmer walking ahead of me, a spry 80 years old, reached down and sliced one off with his machete, offering it up to me in his weathered hand, a gesture of simple generosity. As if on cue, a male Resplendent Quetzal swooped down to a nearby avocado tree, fluttering briefly to snatch a fruit in its yellow bill and then settling on a nearby branch to gulp it down, just a few dozen meters away. It was the best look I'd ever had of the bird, each feather bright and immaculate, colors radiant, its dark eye fixed on mine. Thus, two gifts at once. Although a person might be forgiven for thinking them unrelated, for me they fit as one and spoke volumes about what this supremely beautiful parcel of land had to offer—a bounteous crop and the combination of the wild and the cultivated, together nurturing both people and wildlife.

Such a rich collage is altogether too rare. The secret here, beyond the fact of the farmer's carefully tended fields, was collaboration, in this case between a local wildlife conservation group, an eco-lodge, and a community of farmers. Each brought something of value to the table. The farmers, of course, had land and resources for wildlife; the lodge brought in tourists eager for quality time with quetzals, and willing to spend money to get that; and the conservationists knew how to manage the land for quetzals, retaining nest

and food trees and never crowding the birds. These factors together helped guarantee memorable sightings that other quetzal hotspots often failed to deliver. We'll explore the details of this project, the KABEK project, later in this chapter—it's too good a story to leave out—but for now let's just say that I left there feeling that all was well with the world and that Resplendent Quetzals would still be there for my grandchildren, and probably for their children after them. And with luck they might take home a cabbage as part of their tour!

A month or two later my heady optimism hit a wall, a sobering correction. Wanting to learn about Resplendent Quetzals in other places—beyond Costa Rica's mountains, a region I had come to know quite well—I turned to the scientific literature. Front and center were studies by Sofía Solórzano, the Mexican ecologist we met earlier in this book whose research on the Resplendent Quetzals of Chiapas, in southern Mexico, helped illuminate the links between fruiting trees and the nesting and migration of those birds. But her research went well beyond that. Concerned about Chiapas's dwindling cloud forests, Sofía and her colleagues turned their attention to determining how extensive that loss had been. Their findings were alarming. In just 30 years, from 1970 to 2000, almost 70% of the quetzal forests of Chiapas had disappeared, replaced mostly by coffee and cattle farms but also by small maize farms carved out by families struggling to put food on the table. Projecting similar rates of destruction into the future, Sofía's studies showed that much of the evergreen cloud forest of Chiapas would be gone by 2050, with quetzals surviving in just two protected biosphere reserves.

There was a small silver lining to what I discovered here. First, the two remaining reserves are fairly extensive, together making up more than 200,000 hectares, with extant cloud forest in about 30–40% of each reserve. Protection remains an issue at both sites, even though they are reserves, but the odds seem good that quetzals will hold on there for a while. The second encouraging (and to me utterly fascinating) discovery was how much Sofía and her team had been able to learn from satellite data. A rich trove of details about the earth's surface is to be found in Landsat satellite images, continuously recorded by sophisticated cameras circling the planet day after day, going back as far as the 1970s. Using such high-altitude photos, researchers have been able to track forest loss over time,

This aerial photo documents the destruction of cloud forest in Chiapas, Mexico. About 80% of the cloud forest in this region has been lost over the past 50 years, mostly to cattle and coffee farms, as well as to smaller subsistence plots.

as Sofía did in Chiapas. By focusing on the highland forests most likely to support Resplendent Quetzals, scientists have brought the power of the satellite "eye" to bear on conservation needs in regions as diverse as Honduras, Panama, and Guatemala. This is helpful because for all quetzals, even the Resplendents, we lack reliable data on status and numbers in just about every part of their range. But the satellites give us a way to guess at their status by providing detailed information about the forests that support these birds. The results help us keep a finger on the pulse of quetzal well-being in ways that it would be difficult to do otherwise.

Tropical Forests Today

The sad truth is that tropical forests are under siege, and a growing body of scientific studies reminds us of that virtually every day. In the last two decades alone, about 1 to 2% of the intact primary forests in key Latin American countries where quetzals live (and where we have the best data) have been cut down each year. If that doesn't seem significant, consider that at current rates of destruction at least half the forests of Latin America will fall to chainsaws and bulldozers over the next 50 years. Admittedly, not all these forests are equally vulnerable. Highland forests, those most likely to hold quetzals, generally survive better than those in lowland areas; lack of access, poor soils, and steep terrain slow the rush of humans into such mountainous regions. But increasingly as the years go on, all are vulnerable.

In the face of this we might predict that the *Pharomachrus* quetzals—five species that depend on Latin American forests—are in trouble. Long-term, that's certainly true. But a quick check suggests that, for now, these birds are fairly secure. We see this assessment reflected in official listings. The International Union for the Conservation of Nature (IUCN), a body that tracks the population status of wild animals, has designated all but the Resplendent Quetzal

Female Pavonine Quetzal

as species of "least concern," with the Resplendent considered only a notch more at risk ("near threatened"). Bottom line, in an era of limited conservation dollars, hundreds of other birds need more help than quetzals do.

If we focus on the forests that keep quetzals alive, we find this somewhat optimistic picture reinforced. Take the Pavonine Quetzal, with its immense range throughout the Amazon Basin. Although a distressingly large swath of the Amazon is under siege, it's mostly the fringes that have been decimated so far. By last count, in 2018, there were more than 23,000,000 hectares of intact Amazonian forest in Brazil alone, a vast uncut area with potential to support myriad Pavonines. Similarly, looking at the three Andean quetzals (Crested, Golden-headed, and White-tipped), we see that significant stretches of highland forest remain standing there, despite widespread clearing in many regions. In many ways, it is the remoteness of these forests that offers the greatest protection. Access is often difficult, sometimes impossible, and large portions exist as protected parkland, including Podocarpus National Park, Ecuador, Manú and Rio Abiseo national parks, Peru, and Las Orquídeas National Park, Colombia. So, while it is true that climate change, farming, timber extraction, and mining inevitably result in shrinking forests, in the short term, over the next few decades, the odds seem good that we'll continue to find extensive and viable quetzal forests in the Andean highlands.

We know a bit more about the Resplendent Quetzal's conservation status. We have, for example, fairly reliable estimates of population numbers in a few regions, and information about the status of their forests is increasingly fine-grained. If we paint with broad brush stokes, the picture that emerges for Resplendents looks something like the following. The species, as we've seen, is increasingly threatened in Mexico, where it is now confined to just two protected reserves in Chiapas, Mexico's southwestern-most state. Populations in Costa Rica and Panama appear to be the most secure, thanks to an extensive chain of highland parks and reserves in those countries. In Guatemala and Honduras, however, the picture is cloudier, with habitat increasingly threatened in some areas but with remaining strongholds in parks and in broad reaches of inaccessible highland terrain.

We can look at the conservation of Resplendent Quetzals as an ongoing skirmish among opposing forces. On the negative side are poaching, climate change, and, especially today, forest loss and degradation. On the positive side, we can list protected parks and reserves, spending by eco-tourists that bolsters conservation efforts, and reforestation. In the sections that follow, we look at each of these topics in more detail.

Poaching, the "Nefarious Trade"

It is the nature of things that we rarely have good information on the frequency of a crime. And thus to measure the impact of poaching on quetzals, we're forced to rely mostly on anecdote and educated guesswork. There are few accurate data on the number of quetzals shot or trapped anywhere, with essentially no information for any species other than the Resplendent. In the case of Replendents, we can venture to say that poaching has likely not harmed them to any significant degree, except perhaps in the ancient Aztec era. The population dynamics of this bird are driven far more by habitat—especially by the availability of food and nest sites—than by guns or traps. While it is always troubling to hear about the slaughter of any magnificent creature, from an evolutionary perspective the viability of populations is what matters most, so that losing a few individual birds here and there (or even a few hundred) rarely threatens a species across its range.

History suggests that several populations of Resplendent Quetzals may have had a rough time during the Aztec era, with many thousands shot or trapped each year for their feathers. Apparently most trapped birds were released, so that the impact on populations

The trade in caged birds has had a devastating impact on numerous birds in tropical America, especially those with brilliant plumage and song. Quetzals have been impacted less severely by such trade, in part because they tend to fare poorly in cages.

may not have been as severe as feared, but it's easy to speculate that many of the birds were injured or that their nesting was disrupted as they were stunned or netted, grabbed and handled, and their feathers plucked. Jumping ahead 600 to 700 years, we enter an era when collecting exotic specimens for private and museum markets reached new heights. Here it was the birds themselves, killed and preserved as skins or as stuffed and mounted specimens, that were coveted, not just the feathers. Writings from this time hint that Resplendent Quetzals may have suffered more than other birds, as their dazzling beauty put them at the top of every collector's wish list, and their feathers found their way into women's clothing.

Nigel Collar and others suggest that for several decades at the height of the Victorian era, in the mid- to late 1800s, trade in Resplendent Quetzal feathers for plumes used primarily in women's hats many have reached the level of 800 skins exported annually from Guatemala alone, with perhaps similar numbers in other parts of the bird's range (e.g., Honduras and Costa Rica). Later, in the mid-1900s, records hint that as many as 100 live pairs were being exported per year from Costa Rica for zoos and private aviaries, trade generally overlooked by the government. Largely in response to this, by the late 1970s the Resplendent Quetzal had been placed on the Appendix I list of CITES (Convention on International Trade in Endangered Species of Wild Fauna and Flora), essentially prohibiting the collecting and trade in the species for all but scientific purposes.

Did such harvesting of Resplendent Quetzals threaten the species? Although we have no accurate fix on the overall size of quetzal populations in Guatemala in the 1800s, a conservative guess might be a total of 150,000 to 200,000 birds countrywide (based on the extent of habitat at that time, known population density in prime habitat today, and much smaller human populations back then). If that number is even remotely accurate, this means that less than 1% of Guatemala's quetzals were killed for their feathers in any year, even during the peak trade years of the 1880s. Though killing 800 birds per year may have had serious local impacts, overall Guatemala's Resplendent Quetzal population was never in danger from the trade in specimens. Nor, for the same reasons, were Costa Rica's quetzals put in danger by the collection for zoos of 100 pairs per year.

Alexander Skutch, the early doyen of Resplendent Quetzal studies in Costa Rica, disagreed with this assessment. In the 1940s, he wrote, "The bird was described by some of the early historians of the Conquest; but it soon grew so rare in all the more accessible portions of the Spanish Kingdom of Guatemala that its very existence came to be doubted in Europe, some ornithologists even classing it among the birds of fable. In the 19th century it was rediscovered by Europeans, and soon its skins began to flow across the Atlantic for museums and the cabinets of collectors. This nefarious trade reached such proportions that the Quetzals might well have been exterminated had not so many of them dwelt in wild mountainous regions which even today are most difficult of access and scarcely explored. Most of these trade-skins originated in the Alta Vera Paz in Guatemala."

Although Skutch was a patient and meticulous student of nesting birds, he was also at times a notorious worrywart, prone to exaggeration when it came to anything that threatened wildlife. This was a man known to have hunted snakes in his garden with a machete in order to save the nests of the birds he was studying! There's no indication that Skutch was basing his speculation about the "nefarious trade" in quetzal skins on any published information. It was simply a gut feeling.

A check of museum collections today, the one place that stores reliable data on numbers of quetzals shot during the 19th and 20th centuries, reveals only modest numbers of Resplendent specimens. In North America, for example, about 400 skins of this species are listed in the largest and most accessible bird collections, most of those collected over a period of 20 to 30 years in the early 1900s. That's an average of just 10 to 15 specimens killed per year, a trivial number. In regions where collectors were most active, Resplendent populations likely numbered at least 40,000 to 50,000 pairs in those peak collecting years, so from a regional perspective guns had no real impact on the viability of those populations.

Cloud forest habitat in the mountains of southern Costa Rica

More recently, the occasional potshot by a farmer, or even the more determined hunting of Resplendents for money (the illicit trade in skins), will probably always elude careful study. Such pursuits, by their very nature, are done on the sly. Especially today, when there are significant penalties for shooting quetzals or selling their skins in all regions where the bird is found, we really have no idea how widespread such killing might be. (You won't find quetzal feathers for sale at an airport kiosk!). Fifty years ago, more of the birds might have disappeared into soup pots in remote mountain settlements, with the feathers sold on the side, but there's a lot of remote forest in those regions, guns are expensive (and thus scarce in those impoverished communities), and most farmers have more urgent things to do with their time.

In short, in our lifetimes (or even before us) it seems unlikely that poaching has affected Resplendent Quetzal populations to the point of threatening their survival, even if a few local populations may have been set back briefly.

Forest Loss and Degradation

While news of the shooting of a Resplendent Quetzal may make our heart ache, there is a far more potent and insidious threat to the species today, which is dwindling forests. Some of this started with the Mayans a millennium or two ago, when people cut or degraded over half of Central America's woodlands for agriculture and timber (all, we should note, with primitive obsidian axes and fire). While many of those forests grew back after the Mayans were decimated by conquest and disease, if we look at regions where quetzals are most in trouble today, the issue is almost always habitat destruction.

Although Resplendents managed to dodge forest cutting for decades—the remoteness of their cloud forest haunts discouraged settlement—this has begun to change in the last 50 years as human population growth has accelerated and people have run out of land at lower elevations. Some of this pressure has come from indigenous people, often the descendants of Mayans, whose farming methods (slash and burn) are well suited to that geography. But these people typically produce large families, and their swelling populations have increasingly shifted the farm to forest balance away from intact woodlands. This, along with the economic incentives of growing coffee and cattle, means that Resplendent Quetzals are increasingly threatened by forest loss. Let's look closer at this, region by region.

By way of preamble, we should first note that tropical cloud forests comprised just 0.4% of the global land surface (in 2001), but hold about 15% of total global biodiversity (and are especially abundant with birds, mammals, amphibians, and tree ferns); and even more remarkably, about half of those cloud forest species live there exclusively. So while we look mostly at quetzals, there's a a lot more at stake here as these forests disappear.

Mexico

As we've seen, no other region has lost more forest, with greater consequences for Resplendent Quetzals, than Mexico. The satellite research of Sofía Solorzano and her colleagues, along with their heroic efforts to visit each Chiapas forest that had held quetzals historically (1970s), showed just how extensive the damage has been. Of 39 evergreen cloud forests (ECFs) with known quetzal breeding activity in 1970, 13 had been completely destroyed by 2000, and another 13 whittled down to small remnants, most with extensive damage by cattle. Only five of those remnants, the largest, retained nesting quetzals. Fortunately, Sofía's surveys discovered six new quetzal forests in Chiapas, making a total of 11 ECFs supporting Resplendents in the region (2000). But of these 11, most held only a handful of the birds. And critically, only two of the 11 were (and remain) protected areas: the El Triunfo and La Sepultura biosphere reserves already mentioned, both located in the Sierra Madre mountains of western Chiapas.

Despite this bleak history—and even bleaker outlook—there are glimmers of hope for the forests of Chiapas, and for the quetzals that remain there. El Triunfo is a large reserve; it has almost 120,000 hectares, about 5 to 10% of which consists of the evergreen cloud forests favored by quetzals. A number of NGOs provide funds to hire staff, provide education, patrol the reserve, and monitor wildlife. Although isolated, this quetzal population contains perhaps 2500 to 3000 pairs, large enough to remain viable, at least in the near term. If we assume similar numbers for the slightly larger La Sepultura reserve, Chiapas retains more than enough quetzals to buffer losses over the next several decades.

Nonetheless, conservationists have plenty to worry about. Almost 20,000 people continue to live and farm in El Triunfo. Some illegal poaching and logging continue there, although data on those threats are elusive. Government agencies such as CONANP (the

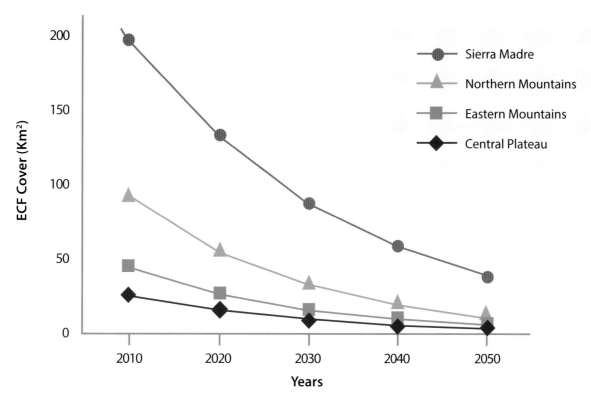

Prepared in 1991, this graph shows projected loss of cloud forest in four key regions of Chiapas, Mexico. Note that only one region, the Sierra Madre, is expected to retain any significant forest cover, owing to two large parks protected there. Of all the cloud forests that Resplendent Quetzals inhabit, those of Chiapas are most at risk. (From Solórzano et al. 2003)

Natural Areas Protection Commission of Mexico) and NGOs, especially FONCET (El Triunfo Conservation Fund), have pioneered conservation efforts in and around the reserve, helping to guide farmers in sustainable practices and buying easements on parcels that most benefit wildlife. While a great start, it's hard to know if these efforts will be enough for the quetzals 30 or 50 years from now. Continued monitoring is key.

Tourism could bring badly needed funds to conservation efforts here, especially if tourism projects are planned with sustainable goals in mind. Today, there are no tourist facilities in the region. Neither El Triunfo nor La Sepultura have a single hotel for eco-tourists, and the few guiding businesses that have gained access to these parks run only sporadic trips. As we'll see, other countries (especially Costa Rica and Guatemala) have built robust businesses around quetzal tourism, bringing significant benefits to the parks that protect them and to the people who live there.

Guatemala

Just south across the border from Chiapas, we enter Guatemala, where images of quetzals adorn the currency and at least a dozen towns include the word quetzal in their name (e.g., Quetzaltenango). Indeed, since the Resplendent Quetzal is the national bird here, embedded in the culture for millennia, one would expect robust efforts to protect quetzal forests. In fact, we find in Guatemala a mix of the conservation failures of Mexico and the successes of Costa Rica (about which more later).

Guatemala suffers from many of the same socioeconomic issues that plague highland Mexico: high rates of poverty; women who lack autonomy and struggle with large families; land tenure policies that lock out the poor; and agriculture that suffers from an over-reliance on cattle and coffee. Yet a few large reserves help keep Resplendent

Quetzals secure in Guatemala, and a small but growing tourist industry promotes the birds in other areas, helping to protect forests that might otherwise fall to the chainsaw.

As seen from a satellite looking down on this country, a single region stands out—the Sierra de las Minas Biosphere Reserve, part of a highland corridor running through Guatemala's mountainous east. This rugged region, a source of precious jade and quetzal feathers for Aztec traders 800 years ago, is today among the largest "protected" forests in Mesoamerica, over 240,000 hectares in all, with about half in cloud forest, 60% of Guatemala's total. While we lack accurate quetzal surveys of most of this region (forest size alone is a prohibiting factor), it's clear that the birds can be found in good numbers in some sections of the reserve, and that the large expanse of prime habitat could hold a lot of quetzals, perhaps as many as 60,000 pairs.

But worrisome trends in land use threaten some of these forests. Nearly 300,000 people, mainly indigenous Q'eqchi Maya, divided into more than 200 communities, live within the boundaries of the reserve. While their main holdings are clustered at altitudes below the cloud forests, the fires they use to clear land often creep up toward highland slopes. In addition, small-scale logging and mining impact higher areas. Beyond that, marble quarrying and damming rivers for hydropower, done at the behest of powerful economic interests with great political clout, continue apace. Like many parks in Latin America, this one suffers from weak government oversight, a lack of funds, and even the activities of bands of narco-traffickers.

Recent studies using satellite data show extensive small-scale logging, mining, and maize agriculture within the reserve; in the last 20 years, a cascade of such disturbances may have had a significant cumulative impact. It is important to have on the ground reconnaissance to assess how these threats are affecting local ecology, and the quetzals nesting there, but even without that method of collecting data, worrying trends are evident. That said, the sheer size of Sierra de las Minas gives some cause for hope, and its rugged, inaccessible terrain helps protect forests and wildlife, at least in the near term.

In addition, a number of organizations have been working tirelessly to protect the cloud forests here. One promising effort has been coordinated by the World Wildlife Fund (WWF) and the local Defensores de la Naturaleza (DdlN). They focus on water. Cloud forests trap a lot of moisture, and the rivers that flow out of Sierra de las Minas nurture lowland farms and industry, economic engines that would flounder without that aquatic lifeline. Educating users about such "ecosystem services," with an eye to voluntary donations, has helped raise limited but badly needed funds for park protection. With more progressive and enlightened backing, particularly from government, taxes levied on water users could funnel critically needed funds back into sustaining Guatemalan cloud forests. Without such help, lowland farms and industry are living on borrowed time.

Resplendent Quetzals have captured the imaginations of Guatemalans for centuries, even lending their name to the currency of that country.

For a look at other ways that quetzal forests in Guatemala have been protected, we step outside of Sierra de las Minas and move north a few dozen kilometers to the region of Alta Verapaz. While cloud forest here is much more limited than in Sierra de las Minas, significant stretches still remain. Within or near protected habitat, especially around the Biotopo del Quetzal reserve, several small lodges have been able to lure in ecotourists and encourage conservation. Earlier satellite data (collected from 1986 to 2000) suggested that deforestation rates were low in Alta Verapaz. More recent data are lacking, although there are hints that pressures are mounting on highland forests in this region. Small maize farms, as well as larger coffee plantations, have been making inroads. It's important to realize how desperate many of these settlers are; highland farms here are tended by people with some of the highest rates of poverty and malnutrition in the Western Hemisphere. These are people struggling to grow food, with limited resources to help them find land. Many who fail are the ones now showing up as refugees at the North American border, desperate for a slice of the American dream.

To meet these challenges, a few conservation groups have been working to cultivate more sustainable farming practices among highland Guatemalans. They teach the importance of crop diversity; protection of surrounding forest; planting organic coffee that is grown in the shade (thus maintaining some trees for birds); and creating healthy soil through the use of mulch and cover crops. All these have clear benefits for wildlife but also for people, resulting in crops with higher yields and that produce a more nutritious diet, as well as more cash when surpluses are sold. Quetzals, as we've seen, may not live full-time on such agricultural lands, but they can often find food and nests along the edges, especially if intact forest is nearby. So far, plantings for quetzals have not been a priority here, but there's clearly potential for such efforts to help concentrate the birds, making them more accessible to eco-tourists and thus boosting spending in local communities.

Honduras, El Salvador, and Nicaragua

The quetzal picture in these countries is fuzzier than what we see in countries to the north and south. Absent sufficient data on quetzal numbers, we can still rely on satellite images to learn a lot about how quetzal forests are faring. especially in Honduras, where forest research has been surprisingly active. An overview of these countries suggests that Resplendent Quetzals are fairly secure in Honduras, despite the social strife and political intrigue that characterize that country. There is a surprising amount of "protected" parkland in Honduras, some of it under siege from encroaching agriculture but little of it critically endangered, until now anyway. Although these parks have skeleton crews and are starved for funds, most continue to function at a minimal level.

Several studies of cloud forest parks in Honduras suggest that core, highland forests, those most likely to support Resplendent Quetzals, are holding their own. Many of these parks are fairly large, and there are dozens of them; in addition, as in so many other quetzal regions, their inaccessible topography slows human settlement. On paper, Honduras has made a significant effort to protect highland forests. That country's 1987 "Cloud Forest Act" protects 37 evergreen cloud forests. Like so many highland parks in Latin America, these tend to consist of a core (exclusionary) zone, surrounded by 1 or 2 lower elevation buffer zones that allow increasing levels of human use. "Sustainable" economic and agricultural activity is encouraged in these lower zones, aided by access to land and credit and by the building of roads. The core zones tend to be at higher elevations, with (on paper) significant restrictions on human use, although those restrictions often go unenforced.

Studies of two different highland parks in Honduras provide interesting snapshots of land use. Satellite data of Celaque National Park from 1987 to 2000 show little change in forest cover, even though indigenous communities of as many as 30 to 40 households were found within both core and buffer zones. All this suggests that human activity is having only limited effects on the quetzals nesting here. A similar assessment emerges from Cusco National Park, another mid-sized Honduran park, this one in the far northwestern reaches of the country. Despite its tiny staff and sparse funds—five rangers in total and an annual budget of just $40,000—its quetzal forests, so far, remain mostly intact.

We can make a pass at estimating Resplendent Quetzal numbers for Honduras, although relying on marginal data likely to be fuzzy. The best estimates come from Santa Barbara National Park, in the mountains of west-central Honduras. Thanks to recent surveys by a birding group in that area (https://www.beaksandpeaks.com), it appears that Santa Barbara's 12,500 hectares hold perhaps 2000 to 2500 pairs. If we assume similarly low density—that is, 20 pairs for every 100 hectares—in 20 of the other largest cloud forest reserves in the country, we come up with a total of about 50,000 Resplendent pairs, likely a conservative estimate for the country as a whole but still a

This tourist lodge in the cloud forest of Honduras allows easy access to viewing birds in the canopy.

number large enough to suggest that Honduran forests support an important reservoir of the northern subspecies *P. m. mocinno*, perhaps a third or more of the remaining population.

Unfortunately, the news about either El Salvador or Nicaragua is not as sanguine. Just a sliver of cloud forest remains in El Salvador, in Montecristo National Park, consisting of a mere 2000 hectares. Anecdotal accounts suggest only a handful of quetzals can be found there now. Much of the country is lowland, with its forests heavily cut, so even in historical terms it's unlikely that El Salvador was ever a stronghold for the species.

Nicaragua holds more promise, although quetzal data for key sections of that country are largely missing. Data largely come from eBird, a robust repository of crowd-sourced birding reports, now worldwide in scope. Northern highland areas such as Matagalpa (Selva Negra) and Dipilto (Nueva Segovia) produce many of the Resplendent Quetzal records. But most forests here are small and quetzal numbers apparently low. With fewer data but far more potential, the vast Bosawas Biosphere Reserve (almost 2.2 million hectares overall, with about 1 to 2% of cloud forest), could hold one of the largest forest reserves in Latin America, perhaps 4000 to 5000 Resplendent pairs. But it's not easy to reach the wilderness highlands there. Trails are primitive or non-existent, and (a new and disturbing wrinkle) land mines planted by rebel groups hiding out in the forest are proving an increasing threat in some regions.

Costa Rica and Panama

From land mines to quetzal lodges—what a difference a few hundred kilometers make! In Costa Rica, we find a country where quetzals occupy a safer niche. My visit in the highlands of Costa Rica with that cabbage and quetzal farmer was a revelation. Throughout the range of the Resplendent Quetzal, this country stands out as a pioneer in conservation. There are good reasons for this, both socioeconomic and political. The Costa Rican ornithologist Gary Stiles helps us understand why. The country, he writes, is "something of a special case with respect to bird conservation in the Neotropics: its stable democratic government, literate public, and relatively high standard of living have made it possible for conservation measures to make great strides here … ." Although threats remain, no other country that is home to Resplendent Quetzals has done more to guarantee a future for the species.

This wasn't always the case. Although 75% of Costa Rica was forested in 1950, over the next 30 years the country lost or degraded nearly two-thirds of its woodlands, owing mostly to government policies that encouraged land settlement and "improvement." Title to land was granted to settlers, often squatters, who then could create farms. Cattle and coffee were particularly favored in this land grab, but small-scale farming (corn, beans, fruit, and vegetables) was part of the picture too. While high altitude quetzal forests tended to fare better in those years than lowland forests, thanks as always to their inaccessibility, they were not immune to cutting. Cattle proved especially easy to move up onto those slopes, requiring only patchy clearing to establish pasture, and minimal care once established.

But with changes in government, and growing environmental awareness, a shift in Costa Rican land policies emerged during the 1980s. Science was beginning to reveal the extraordinary biodiversity to be found in the country, especially in forests. Largely in response to this awakening, nature tourism began to blossom here, with robust revenue flowing from that. Responding to all this, the Costa Rican government swung into action. By the early 1990s, almost 20% of the country was under park protection, a significant accomplishment even in global terms. Although Costa Rican parks protected a variety of regions and ecosystems, in many ways highland forests benefitted the most from these efforts. One could argue that Resplendent Quetzals served as an emblem of sorts in this effort—they were often featured in promotional literature—and of course they stood to gain as cloud forests were set aside.

The jewel in the crown of Costa Rica's emerging parks was La Amistad International Park, protecting almost 400,000 hectares in southern Costa Rica and northwestern Panama, with roughly 30 to 40% of that in highland forest, prime quetzal habitat. The splendid Talamanca Mountains run down the center of the park, which is bordered on the north by a constellation of smaller parks and reserves (e.g., Chirripó, Rio Macho, Tayni, Quetzales National Park) and to the south, in Panama, by a smaller but still significant mix of protected areas. This impressive stretch of intact forest, over 200 kilometers long, now holds nearly the entire population of *P. m. costaricensis*, the southern race of the Resplendent Quetzal. Because this habitat is so pristine, it has special value for ensuring the long-term prospects of quetzals. Northern Resplendents tend to exist in smaller and more isolated populations, limiting the genetic mixing that helps maintain vigor to in a species.

Although there are persistent threats of forest loss in Amistad, due mostly to agriculture on indigenous reserves within the park or along its edges, they appear to be relatively minor. Satellite data from the Panama side of the park confirm just 0.6% annual loss of forest there in recent years (1998 to 2018), and such loss seems to be compensated for by regenerating forest.

That said, park management in Panama, and thus the future of quetzals there, is less well-regulated than it is in Costa Rica. Lax government oversight has been an issue in some regions, allowing farms to expand onto sites that are under the park umbrella. On the positive side, there is a small but active tourist industry focused on

A cloud forest stream in the southern range of the Resplendent Quetzal, near Boquete, Panama.

quetzals in highland Panama, especially around the towns of Boquete and Cerro Punta. This may help slow the degradation of parkland. Quetzal density can be impressive in some regions there, with up to 50 pairs per hectare. It remains to be seen how well these southernmost Resplendents will fare in the decades ahead.

Climate Change

Year after year of record heat and drought-sparked wildfires along the California coast set against unprecedented winter snowfall in the nearby Sierra mountains; mile after mile of Australian and Caribbean reefs dying from overheated seas, while French wine grapes are zapped by rare spring frosts; Greenland glaciers melting at ten times historic rates, but Texas airports closed by snow for the first time in decades … climate "weirding" begins to seem like a better term for what the planet is experiencing. But the facts—the ticking time bombs nestled up against the ecological foundations of our living systems—are inescapable. Although there's still a lot we don't know, it's clear that our world is heating up fast, faster than ever before. And for many species, including ourselves, the consequences are likely to change just about everything about our lives.

Quetzals live in cool, cloud-bathed habitats, which you might expect would shield them from the worst of global warming. But heat (and likely drying) will be inescapable even at those elevations, and the birds are also confined to narrow altitudinal bands of forest, so they occupy a restricted niche. As their mountain world gets warmer or drier—and as the trees they rely on dwindle or shift in range—will they be able to find refuge in new and welcoming places? Fairly robust predictions are telling us that by the year 2100 Costa Rica and much of Central America will likely experience average temperatures at least 2 to 3 °C (3.6 to 5.4 °F) higher than those recorded today. That's a huge jump, one that could make day-to-day life almost unbearable for people who live in humid lowland areas, but could also change quetzal forests in ways we can only begin to predict.

Compared to the information we have for most birds living in tropical highlands, the well-studied ecology of Resplendent Quetzals gives scientists a better foundation from which to peer ahead at the impacts of climate change. Studies show trees that have evolved in cool montane habitats do poorly as temperatures rise, so the only surviving remnants are likely to be found farther upslope as their world warms. But temperatures are likely to rise so quickly in the next one hundred years and beyond that many trees may not be able to shift ranges fast enough to keep up. Stepping back in time, a few thousand (or even a few 100) years ago, shifts in forest composition seemed able to track slower changes in temperature or rainfall more easily, with quetzals and other animals moving along with those shifts and maintaining viable populations.

Today, the odds of that happening look slim. Scientists at Purdue University studied projected changes in bird communities in the Tilarán Mountains (the Monteverde Preserve, Costa Rica) during the 21st century, with a focus on Resplendent Quetzals. Based on two different models, one hinged to projected changes in temperature, the other to changes in precipitation (less rain, especially in the dry season), the study suggests that by 2100 Monteverde quetzals will be alive and reproducing but living in a smaller, more restricted, range with lower numbers. How much lower depends on the model selected, but losses could be as high as 80 to 90%. In a nutshell, it seems likely that at least half of Monteverde's quetzals will be gone in a century, with the remainder living at higher altitudes, where forests may be able to retain the trees, especially the Lauraceae, whose fruits these birds depend on. For individuals needing to move beyond such core breeding areas outside the breeding season, as many Resplendent Quetzals do, the challenges are likely to prove even greater.

To refine our projections, we must keep in mind the feeding ecology of these birds. Even though there are more than 30 species of Lauraceae with fruits that Resplendent Quetzals are known to eat, just three or four of these make up the bulk of their diet at any one time, especially during the critical period when young are being fed at the nest. How each of these trees will respond to drier and warmer climates is anyone's guess. And it isn't at all clear that Monteverde forests provide an especially good model for determining the fate of this quetzal across its range. Indeed, there is reason to think the bird may do better in the Talamanaca Mountains to the south, where elevations are higher and the forest more extensive, factors that could help buffer the species from the worst of warming and drying trends. The same may be true in the large reserves of Guatemala and Honduras.

In Mexico, where quetzals are increasingly under siege from deforestation, climate change is likely to bring significant new burdens for these birds. Recent studies show just how vulnerable their montane cloud forests are, owing especially to their narrow and fragmented distribution. Sadly, it seems that many such "islands in the sky" face a bleak future. Based on predicted shifts in temperature and rainfall, 68% of these forests may vanish by 2080, with 90% of Mexican cloud forest that is currently under protection likely to become highly altered by then, with vegetation communities very different from what we see today.

Stepping back, it's probably fair to say that quetzal cloud forests are among the terrestrial ecosystems most vulnerable to the impacts of short-term climate change. Accelerating threats to Resplendent Quetzals are likely to impact already compromised populations especially hard, probably eliminating many of them. Those least threatened today may survive the increasing heat and dryness in refuges at higher elevations, but there's still a lot we don't know about what those forests will look like in 100 years, or how long those refuges will last after that. Human intervention could help; one possibility is to move (by human hand) Resplendent Quetzals from areas where they are most at risk to higher elevations, where they can find refuge in forests that retain food and nest sites.

Similar worries exist for quetzal species that depend on Andean highlands, where some of the best modeling suggests that forests will struggle to keep up with projected changes in rainfall and spikes in temperature. Even though mountains are higher there than in Central America, with more potential buffer at altitude, it seems the changes may happen too fast for forests and their quetzals to keep up. In addition, poorly understood biological forces may prevent the upper edges of these tropical cloud forests, the tree line, from moving onto what are currently mostly grass and shrub lands, known as páramo. This so-called "grass ceiling" is perhaps yet another factor

The Yellow-throated Toucan (*Rhamphastos sulfuratus*) is a predator of Resplendent Quetzal nests, especially in the southern parts of this quetzal's range.

that will squeeze highland quetzals into smaller and smaller bands of forest as the planet warms.

A final wrinkle in the climate change scenario has to do with predators. Warming temperatures and changing forests appear to be expanding the ranges of some species of birds known to rob the nests of Resplendent Quetzals. Particularly worrisome is the Yellow-throated Toucan (*Ramphastos sulfuratus*), whose long and dexterous beak lets it reach into quetzal nest holes and seize eggs and young. Anecdotal evidence from Monteverde, Costa Rica, suggests that toucans, whose core range is lowland forests, may be moving upslope more regularly, and staying longer, putting nesting quetzals at risk. Higher elevation quetzal populations in other parts of Costa Rica seem to be escaping such intrusions, but this trend merits continued monitoring; perhaps creative conservation minds can find ways to make quetzal nests less vulnerable to toucans and other predators.

Tourism

Quetzal tourism is something of a mixed blessing, though on balance it has generally proven beneficial for the species. Almost 100% of this tourism is centered on Resplendent Quetzals in Central America, with Costa Rica and to a lesser extent Guatemala attracting the lion's share of tourists. I like to joke that this quetzal has become the "million dollar bird," and in fact it is proving to be all that and more.

This roadside billboard near the southern Costa Rican town of San Vito indicates how important Resplendent Quetzals are to tourism, though from this spot one must travel another 20 to 30 kilometers to higher elevations in order to find the species reliably.

Humans have been fascinated by quetzals since the Aztecs and their Mayan and Olmec ancestors before that, although it seems that they were more interested in the bird's feathers than the bird itself. Today, attention has shifted to the birds. A glimpse of a male Resplendent in the wild, with its long, shimmering tail plumes swaying in the wind, or a female at a nest hole, dangling a lizard in her bill, is often the crowning achievement of a nature tourist's cloud forest experience. For birders and naturalists, many of whom travel thousands of miles and spend thousands of dollars to see it, this is the animal that's at the top of their list. Driving roads in Costa Rican or Guatemalan quetzal country, images of Resplendent Quetzals adorn billboards and plaques, and the bird has become part of the logo identity of dozens of B&Bs, restaurants, and (in one tiny Costa Rican village I visited) even an establishment selling car parts!

Interestingly, quetzal tourism often thrives in places where the species was once under threat. All of which brings us back to an obvious fact—the value of conserving forests. Nowhere is this more evident than in Monteverde, a community in the Tilarán Mountains that was founded by Quakers. Here, during the 1970s, motivated citizens led by quetzal researcher George Powell raised money to set aside more than 10,000 hectares of intact cloud forest, with significant parcels added over the following decades. Today the Monteverde Cloud Forest Preserve pulls in over 70,000 visitors a year, a source of valuable income for the entire community. Rough, back-of-the-envelope calculations suggest that the quetzals, a main attraction here, have been "earning their keep." Total annual income from tourists in Monteverde, divided by the estimated number of quetzals in the reserve (300 to 400 pairs), gives a "value" for each bird of roughly $12,500 per year! While such accounting may strike some as vulgar, the numbers are telling. Obviously, quetzals aren't the only draw here, but they're certainly front and center, a forest icon, helping to convince even the most entrenched skeptics that preserving cloud forests and their quetzals can, in the right places and with the right effort, make a real difference for local economies.

A small research center at the Monteverde Preserve has gathered valuable data over the years to help guide management of quetzals and their forests. Some earlier research projects that helped launch this center, or were satellites to it, include Nat Wheelwright's detailed work on Resplendent Quetzal diet and feeding ecology and George Powell

Right: Here, Don Wuillain Selano, a farmer who plays a key role in the Kabek pro-Quetzal community in Costa Rica, displays shed Resplendent tail feathers that he has collected on his property, an evident source of pride.

and Robin Björk's pioneering use of radio transmitters, which showed how mobile these quetzals are outside the breeding season. Tourist fees at the preserve have helped bolster some of this work, and today contribute significantly to the budget of Costa Rica's Tropical Science Center (based in San Jose), which oversees the Cloud Forest Reserve. Critics argue that more of the gate fees should stay in the community; there may be ways to remedy this in the future.

While such tourism-funded conservation got a start in Monteverde, it has since taken on new life in other parts of Costa Rica, especially in the central mountain region, where Paraiso Quetzal (PQ), an inspired eco-lodge with a focus on quetzal watching and photography, has joined forces with one of Costa Rica's most effective conservation organizations, the Costa Rica Wildlife Foundation (CRWF). This collaboration has fostered a local, community-wide venture, now a model for tourist-based quetzal conservation, and one that introduced me to that generous cabbage and quetzal farmer I met in my early visit to Costa Rica. The elements of this success have been remarkably simple but effective.

As elsewhere, quetzals in the central mountain region move up and down slope depending on seasonal shifts in food availability, often leaving protected forest areas at least temporarily for more open landscapes, generally small farms. The good news is that many of these farms have retained trees and have at least some protected forests nearby that have escaped degradation. As long as *aguacatillo* trees remain intact and productive, and the forests have a reservoir of breeding quetzals, the birds continue to visit these settled areas and to thrive in this habitat mosaic. By encouraging farmers to keep the *aguacatillos* (and plant new ones), and by linking PQ birders and photographers with farms where the quetzals are most active, the lodge can deliver a top-notch experience for guests, while the farmers are compensated for the role they play. This effort has become known as KABEK conservation.

Dozens of farm families are now on board with KABEK, many of them seeing a significant boost in income thanks to tourist dollars. I like to think of them as quetzal farmers, or at least farmers with a quetzal side crop. Yes, they also raise cabbage, potatoes, broccoli, tomatillos, and calla lilies, among other crops, but they earn just as much, or more, by ensuring that quetzals find welcoming habitat on their farms. Many of these people have limited incomes, so the quetzal dollars make a big difference. Perhaps just as important,

Left: Lake Atitlán, Guatemala. Remnant cloud forests in the background peaks seen here hold small numbers of Resplendent Quetzals, a boost to the eco-tourist industry in this region.

being part of this quetzal network is a source of pride for these locals. Tourists, especially photographers, come from around the world to visit. In any one week, a KABEK farmer might welcome a photographer from Japan, a keen birding couple from the UK, and a young family of aspiring naturalists from Quebec. How many of us can boast the same diversity of visitors to our own neighborhood, however grand or humble it may be?

In Guatemala, we find a handful of quetzal reserves, with at least a few lodges that cater to tourists wanting to see quetzals, but all on a much smaller scale than in Costa Rica, and with fewer links to conservation. That said, overall, tourism is robust in the country; Guatemala currently welcomes over two million tourists a year. It's not clear how many visit to see nature, or quetzals in particular, but if even 2 to 3% spend time on a quetzal tour, that's 50,000 people annually bolstering local economies and helping to keep cloud forests alive. A few regions stand out: Biotopo del Quetzal and Refugio del Quetzal, both in the municipal protected area of San Rafael de la Cuesta, welcome a modest but steady stream of quetzal tourists each year, with Ranchitos del Quetzal, a lodge nearby, catering to such groups and adding to the mix with its own small but vibrant protected area. Farther west, on the Pacific slope, small preserves such as Atitlán Volcano boast a few quetzals as well as other natural wonders, and the visitor response has been impressive. What's missing in all these places, after what we've seen in Costa Rica, are efforts to integrate such tourist flows into a larger conservation scheme that bolsters a research community. Guatemala needs a vibrant tropical science center; quetzal tourism, if further developed there, could help launch and sustain that.

Finally, we must consider the less salutatory sides of tourism. I remember my first quetzal tour in the San Gerardo de Dota region of highland Costa Rica. My companion and I had rented a small house there, near a spot known for its Resplendent Quetzals—much of the Dota, a lovely area still heavily forested (known by some as the Switzerland of Costa Rica), is home to these birds. Up before dawn, we drove seemingly remote gravel roads, in route to a rendezvous with our guide. We found him, alas, alongside dozens of cars parked by the road, where several large tour buses disgorged an endless stream of sleepy high school students. Our guide had set up his telescope on a male Resplendent high in a distant tree (the only quetzal we saw that day) and the students were dutifully lining up for quick looks before they disappeared back to the bus for a snooze or time on their smartphones. By the time my turn at the telescope arrived, the bird had disappeared. So much for the quintessential wilderness experience tracking Latin America's iconic cloud forest bird!

I've had somewhat similar experiences in Monteverde, Costa Rica, where the cloud forest reserve there often fills with tourists less than an hour after opening. To its credit, the reserve limits the number of visitors allowed in at any one time. Nevertheless, if late, you stand in line until space opens for your group; if early and you are allowed in, you are often within sight or earshot of other visitors much of the time. Perhaps not surprisingly, quetzals are often hard to find there.

It doesn't have to be this way. It's possible to avoid the commercialization of the quetzal experience, to preserve the magic of one's first look at streaming, iridescent tail feathers high in the canopy. The way to do this is to find less traveled routes, including the southern Talamanca Mountains in Costa Rica; reserves in the Alta Verapaz, Cerro el Amay (Quiché), and Sierra de las Minas regions of Guatemala; and Santa Barbara National Park in Honduras, among other highland parks there. Easier said than done, perhaps. Unfortunately, Latin American (and especially Costa Rican) parks often lack facilities and access. The emphasis is on preserving biodiversity, less on tourism, so it can be hard to find satisfying quetzal viewing sites. Nonetheless, park edges (often accessible by road) can provide excellent opportunities for seeing the birds; at the right season, particularly early in breeding, it is often possible to see impressive numbers of quetzals in just a few hours.

Thus, while tourist crowds may boost local cash flows, they can detract from the wilderness experience that most quetzal seekers wish for. In addition, there is the threat of disturbance to the birds themselves. Quetzals are most vulnerable to disturbance when nesting. Although Resplendent Quetzals can be quite tolerant of people once they have committed to a nest hole and laid their eggs, discretion is still the rule. Keeping groups small and quiet; confining photographers to blinds and avoiding camera flash; restricting visits during the busiest periods of the day for the birds—all these precautions help ensure that a nesting pair isn't compromised when raising young.

Standing back, what can we say overall about this relatively new phenomenon of quetzal tourism? Clearly it shows evident promise as an anchor for forest preservation. We find in Resplendent Quetzals an unforgettable forest icon, capable of drawing tourists from thousands of miles away. And these people arrive with money in their pocket. In the communities that stand to benefit most from this spending, a little bit can go a long way. That's good news for conservation. While controlling the tourist flow may increasingly become a challenge in some places, opening up new regions for quetzal viewing could take the pressure off others. Costa Rica leads the way in making quetzals a tourist draw, and projects in Monteverde and with KABEK are models for what's needed in Guatemala, Mexico, and Honduras. But even in Costa Rica, one could argue that quetzal tourism is still a fledgling operation. Only two or three regions promote it with any success, and (taking cues from wildlife viewing in other parts of the world) there are many ways to enhance the tourist's experience.

Restoring Forests

Cloud forest, the primary habitat for the Resplendent Quetzal and its Andean highland relatives, composes just 1 to 2% of Latin America's total tropical forest. Yet about half of such cloud forests have been cut to date. That's the bad news. The good news is that reforestation projects are growing in scope and number. In at least a handful of regions, these efforts are likely to benefit quetzals in the long run. And, perhaps most importantly, such pioneering reforestation projects could become a model for restoring habitat in regions that currently lack the resources or initiative to begin work.

Restoring forests is an inherently satisfying endeavor. Imagine turning an abandoned cattle pasture or coffee farm into vibrant and species-rich woodlands. Until now, the process of doing so has involved as much art as science, but increasingly, in a few regions (especially Costa Rica and Ecuador), we are seeing strong, fruitful collaborations that have begun to perfect the intricacies of the process. The main challenge with renewing forests on degraded land is getting seeds to the right locations and having them sprout and grow beyond the critical early seedling stage. Few seeds travel far from the parent tree; dispersal is a challenging issue in forest restoration. One solution is to start the seeds in protected nurseries and then hand-plant the seedlings in prime habitat, where they tend to grow quickly given a salubrious tropical climate. While labor intensive, the rewards can be great because fast growing seedlings shade out tenacious grasses and ferns that are a major barrier to getting forests jump-started on their own. As shade trees develop, new shade-loving trees move in under them, and suddenly what was a hot and scrubby pasture begins to look like a real forest.

We should say, however, that no one has recreated a quetzal cloud forest from scratch—the complexities are just too great to

Right: An adult male Resplendent Quetzal perches in pristine cloud forest habitat, Las Tablas Protected Zone in southern Costa Rica. Few areas provide better looks at this species during its nesting season.

achieve over a short period of time. And yet we must try. In highland Ecuador, efforts to restore forests have produced woodlands that, after 20 to 30 years, are less diverse than primary forests, with fewer rare and endemic trees, but nonetheless count as healthy forests with a closed canopy and increasingly diverse wildlife. If we were to just focus on quetzals, one wonders what we might accomplish. If their Lauraceae food trees were made a priority in replanting, and if nest boxes were installed to make up for the lack of decaying trunks with nest holes in them, perhaps these birds could recolonize younger forests more quickly than they'd be able to do otherwise. Without such efforts, it's likely to take almost a century, and perhaps more realistically 200 years, to establish the woodlands that quetzals would naturally colonize and thrive in.

Trends suggest there may be plenty of such land to experiment with. Humans in Latin America are abandoning montane agricultural land at accelerating rates, owing to depleted soils and to the changing sociology and economics of the people living there. In Colombia, over 2 million people have left rural areas for cities in the past two decades, displaced by civil unrest and by the lure of better jobs and a more stable life in urban settings. Many of these people are coming out of montane areas and there are fewer and fewer to take their places. Such trends are less evident in more politically stable countries such as Costa Rica and Panama, especially given their government-subsidized indigenous populations, but even there urban areas are becoming increasingly attractive to a jobs-hungry populace.

From an ecological perspective, the aim of tropical forest restoration is to accelerate or "jump-start" succession, helping forests recover lost ecosystem functions and species diversity more quickly than if left to regenerate unassisted. We can't leave our discussion of reforestation without mentioning an ambitious project taking shape in Costa Rica, Osa Conservation's (OC) Reef to Ridge initiative. OC is based on Costa Rica's Osa Peninsula, a center of lowland forest diversity that includes world-renowned Corcovado and Piedras Blancas national parks, as well as marine reserves in the adjacent Golfo Dulce. Creating forested corridors that link these biologically rich lowland areas to the equally rich forests of the Talamanaca Mountains, 40 to 50 kilometers away and at least 3000 meters higher, is the goal OC has set for itself. The scale of the project is impressive. Hundreds of thousands of trees are to be planted; hundreds (if not thousands) of land use easements must be negotiated (much of the work will be on private land); and years must be devoted to monitoring forest growth and wildlife activity to determine the success of the project and what corrections are needed as it moves along.

Quetzals have less to gain from this effort than many other species do, species such as jaguars and peccaries that wander farther and depend more directly on lowland forest. But regenerating forests on the lower flanks of the Talamanca Mountains, a region that has been especially degraded by cattle, can't help but benefit Resplendent Quetzals as they move out of cloud forests in the non-breeding season, and head to lower elevations looking for food. Efforts to integrate Lauracae avocado trees into the mix of planted species could only help here.

Such efforts help turn back years of forest loss and degradation. Costa Rica has been gaining forest at a remarkable clip, restoring and protecting more than 20% of its formerly degraded forests in the last few decades; no other country in Mesoamerica has even come close. But this is a "scalable" effort, ideally. By proving, in Costa Rica, the concept of reforesting on a grand scale, ecosystem to ecosystem, via NGO and government partnerships, the country creates a blueprint for other countries. Not surprisingly, large scale reforestation faces challenges, among them government red tape, funding, and murky land titles. Overall, however, Costa Rica has shown how significant forest areas can be secured and restored, with quetzals and other wildlife the likely beneficiaries.

To conclude on a mainly positive note, surveys of Resplendent Quetzals have indicated populations that are more robust than earlier estimates had suggested, owing mostly to remarkably high densities in many areas. Plus, some surprisingly large forest reserves that still remain, especially in Costa Rica but also in Guatemala, Honduras, and even Chiapas, Mexico, where the species probably faces its greatest challenges. In other regions, for other quetzal species, we often lack the information we need to assess populations, although there are a few large Andean reserves that harbor significant numbers of quetzals. If the destruction of forests is the biggest current threat, equally pernicious is climate change with alarming trends (heat and drying) that threaten large swaths of the forests that quetzals depend on. Within a century, we could lose half to two-thirds of the quetzals we see today. But we aren't obligated to remain passive witnesses of the destruction, and our final chapter outlines plans of action.

Right: A close-up of a male Resplendent Quetzal highlights the iridescent sheen on the plumage of this magnificent bird.

LOOKING AHEAD

A need for sound ecological data as a prerequisite for … effective conservation measures [as] reflected in the apparent inadequacies of safeguards put in place to protect the Resplendent Quetzal … .

J. M. Forshaw, *Trogons: A Natural History of the Trogonidae, 2009*

Trail marker in the Santa Elena Cloud Forest Reserve, Monteverde, Costa Rica. Resplendent Quetzals have an outsized role in attracting tourists to Monteverde.

On days when my imagination plays fast and loose with me, I sometimes let it wander in the direction of fantasy financing. We have all imagined these scenarios—the lottery ticket bought on impulse that comes up a winner; the stock pick that hits, beyond your wildest dreams; or the small business you've nurtured for years finally finds a lucrative niche, and buyers materialize at the door. In my world, the world of conservation, I fantasize about an encounter with the perfect Maecenas. What if a generous donor heard my quetzal tales and in response handed over a significant amount of money, saying "here, spend it wisely … just be sure whatever you do will make a difference, a big difference on down the line for those magnificent birds."

It's a fun challenge to contemplate. But where would one begin? The biologists will tell us to start with research, while conservationists would argue that habitat is being lost so fast that we must save it where the threat is immediate and then ask those research questions later. Both are probably right. Targeting conservation dollars wisely means having access to key data such as baseline population numbers and indicators of forest health—letting us identify which populations are most at risk—and using such information to prioritize conservation on a large scale. But we also need to know what drives natural fluctuations of quetzal numbers and whether habitat management can reduce the dips to compensate for factors that cannot be controlled or ameliorated.

To argue their case, the biologists will remind us how little research has actually been done on quetzals, especially the South American species. Even for the better known Resplendent Quetzal, basic aspects of ecology and behavior remain unknown: how long individuals live; causes of mortality; how faithful individuals are to mates and nest sites; what influences breeding success beyond food abundance; why some individuals migrate after breeding while others stay put (and how migration patterns vary year to year and region to region); and how far young move after fledging, before finding a more permanent home. All these questions, and more, remain unanswered, or have not been addressed fully. Yet without such information, it is hard to know how best to allocate conservation funds.

Refining Population Estimates

Getting accurate estimates of quetzal numbers remains a top priority. Without knowing which populations are limited and at risk, we'll never be able to send our research and conservation funds to

Left: Disappearing quetzal act. Here we see a male Resplendent exploring its nest hole in Chiapas, Mexico. With concerted conservation efforts, sights like this one will be available for decades to come in the cloud forests of Chiapas and elsewhere in Mesoamerica.

where they're needed most. Until now most population estimates have been based on limited data. We need to do better, and, increasingly, we can.

One way to collect better data relies on acoustic monitoring. Because quetzals tend to be vocal, at least during the breeding season, sensitive recording devices installed in prime habitat and left for months can yield impressive results. Using computers, such recordings can be categorized by time and location, allowing researchers to draw a more detailed picture of where quetzals are found—and in what numbers. Installing an array of these devices in priority cloud forest habitat brings regional estimates of density within reach. Combined with occasional ground surveys (well-trained human eyes and ears are needed for back up), a small team of researchers can harvest density data that would take months to achieve by eye and ear alone—and would likely be far less accurate.

Once we have density estimates for a limited area, we still need to know how much habitat is available for a given species, starting with regional populations. Although the raw satellite data needed to answer such questions are increasingly available and sophisticated, processing those data and making sense of them has rarely been a priority in quetzal research. We could change that. A small team of computer-savvy biologists could, in a few months' time, uncover much of the information we need to estimate forest size in regions where quetzals are living today. How much cloud forest remains in Honduras,

A Plate-billed Mountain-Toucan from the western slopes of the Andes, Ecuador. Smaller toucans like this one are known predators of quetzal eggs and nestlings.

for example, and how much of that matches the preferred elevational range of the Resplendent Quetzal? Just as important as quantity is quality: satellite data can also shed light on the degree to which standing forests have been damaged by by cutting, the incursions of cattle, maize farms, and the like. All this would make a worthwhile thesis project for an ambitious graduate student or two, especially if they were to expand the scope of the project to include regions and countries that remain poorly known.

Tracking Movements

As we have seen, in some areas Resplendent Quetzals tend to leave nesting areas at the end of each breeding season, often heading to lower elevations, where trees are on different fruiting schedules and food is more available. Yet few of these birds have been tracked; those that were studied were tracked in just a single area (Monteverde, Costa Rica), and much of that research was carried out 30 or more years ago. To confound matters, anecdotal reports suggest that in some years many Resplendents remain on or near breeding areas, apparently finding adequate food without having to "migrate." Knowing if these birds move—and when—helps target habitat protection for the full range of this species. Data for Andean quetzals are also essential, as forest clearing is increasingly transforming many of those slopes.

The good news is that we now possess very effective technology to track these birds. Initial efforts with Resplendents relied on radio-telemetry, with lightweight radio transmitters attached to individual birds. The challenges of following transmitting birds on foot (or even by plane) are legion, but there is now a more efficient tracking technique that relies on an array of stationary, automated recording stations. Each station holds a powerful receiver capable of detecting a signal at least 10–15 kilometers distant; thus, a group of stations can cover a broad area. Known as the Motus Wildlife Tracking System (MOTUS) tracking system, this is an international collaborative research network able to track a variety of wildlife, from large birds to dragonflies. Cloud forests have yet to be targeted to any significant extent, so there is potential for quetzal researchers to lead the way here. With a network of MOTUS receivers installed in prime quetzal habitat, and birds fitted with tracking radios, location data would be available 24/7, with each individual identified by its unique tag signature.

Male Crested Quetzal, Peru. This close relative of the Resplendent, with its bright red eyes, is an Andean cloud forest icon.

Quetzal researchers using MOTUS tracking could learn fascinating details about the continuity of migration movements, telling us if the same individuals migrate each year and whether they travel to the same locations; whether the birds move as individuals or as families or groups, simultaneously giving us data about how long young birds stay with their parents and where they go once independent. In addition, MOTUS stations would let us harvest data on geographic variation, revealing whether patterns of Resplendent Quetzal movement in north-central Costa Rica are typical for the species, or if populations elsewhere behave differently.

Marking Birds for Life

Although Alexander Skutch's ground-breaking study of nesting Resplendents in Costa Rica tells us much about how these birds go about getting a new generation on the wing, we lack all but the barest sketches of information for other nesting quetzals. Even for Resplendents, one key piece of data for understanding life histories is still missing, which is the identification of individual birds. Without knowing who's who—if you are seeing the same bird week to week or year to year—researchers cannot begin to understand population dynamics. We'd like to know how long individuals live; how often they change mates and nest sites; and how many young an individual or a pair can raise in a lifetime.

For other birds, the ability to record data for specific individuals got a big boost decades ago with the development of increasingly sophisticated leg bands, markers with unique color combinations or numeric coding that enabled observers to identify each banded bird. Quetzals take poorly to bands, however. Their short, feathered legs are generally too small to accommodate them, thus the need for other methods.

One such option is the PIT tag (Passive Integrated Transponder), a microchip that can be attached unobtrusively to a bird, or even inserted under its skin. This same technology is now used widely for human pets; dogs, cats, and even parrots, for example, are surgically "micro-chipped" with a subcutaneous tag they wear for life. Each chip has a unique code readable by an electronic device, a transceiver. In many ways these tags function as modern, sophisticated bird bands, and they have been used in hundreds of research projects involving tens of thousands of wild birds. The only limitation to the technique is that tag and transceiver must be in close proximity to register. This makes the PIT tags particularly well suited for use around nests or other locations (e.g., perches) to which an individual bird returns repeatedly.

For quetzals, nest holes are an ideal location for a PIT tag reader. Given the low cost of this technology, dozens of holes could be fitted with a transceiver, and tagged birds identified each time they come and go from the nest. Because the tags last for the lifetime of a bird, researchers can identify individuals year after year, lending continuity to quetzal studies that has been missing so far.

PIT technology could answer a host of other questions. How faithful, for example, are quetzals to nest sites? Within a breeding season, how often do they move to a new site to raise a second brood? In the following year, do they reuse old nest holes or move to new ones? It would also help conservationists know whether building artificial nests increases productivity and perhaps reduces territory size.

There's still a lot we don't know about what makes a quetzal choose a particular nest. Having records of which holes get the most traffic could help answer this question, and more broadly help us understand if the lack of nest sites is limiting populations of these birds.

Lastly, tagging young before they leave the nest could help us understand many aspects of longevity and reproductive behavior. How long it takes an individual to find a mate, for example, and how often these birds change mates; and which individuals are most successful in raising young. Once a local population is tagged and readers installed at nest holes, year-to-year studies become routine, especially if new fledglings are tagged each year. A modest project beginning with, say, three

or four local Resplendent populations (each with perhaps 20–30 tagged birds), studied over a decade, would go a long way toward boosting our understanding of key aspects of the life history of this iconic species.

Seeing Life at the Nest

Webcams (video cameras installed at nests that stream images to the Internet) now let thousands of fascinated viewers see details of nest life for a wide variety of birds, details that previously were available to

A male Resplendent Quetzal feeds a laurel fruit (*aguacatillo*) to its young at a nest hole. Male Resplendents take an increasingly active role in caring for nestlings as they age.

only the most dedicated researchers. Webcams have been especially revealing with hole-nesting birds, where eggs and nestlings are hidden from view. Such cameras have the potential to open up quetzal nesting to the world and spark interest in people who have never seen a quetzal and have little chance of ever visiting a quetzal nest.

There is a particular magic in seeing eggs being laid and incubated, the parents taking turns at the nest; in watching those eggs hatch and the emerging nestlings fed and brooded; and in witnessing the young grow feathers and interact with their siblings. All this does more than appease our curiosity. For scientists, there is the chance to record useful data (diet, feeding rates, nestling behavior), much of which has so far been unavailable—and in the case of quetzals, all hidden from view inside a hollow tree. Webcams also allow scientists to recruit viewers as "citizen scientists," a network of amateurs trained to record information that cumulatively proves useful to science.

Opening Resplendent nesting to a web-based audience, including students in classrooms, citizens throughout a country, tourists at eco-lodges, and keen birders everywhere could broaden the quetzal audience beyond the lucky few who are able to travel and see the birds in the wild. Viewers become potential stakeholders in a species and its supporting habitat, guardians of its future, and donors to scientific research and conservation. While the initial webcam investment can be expensive, the rewards have great value, with conservation the ultimate beneficiary.

Ideally, we would already have in hand the results of many research projects before outlining a sound conservation strategy for quetzals, but, as the saying goes, "you go to war with the army you have." That said, we do have sufficient information today about Resplendents to chart out a conservation plan, a plan that will surely be refined as our understanding continues to develop.

Nest Boxes

As we've seen, nest-hunting Resplendents and all other *Pharomachrus* look for dead trees at the just the right stage of decay, with access to interior hollows via one or more woodpecker holes. Such exacting requirements suggest that nest sites may be limited for this species, as has been shown for parrots and other tropical hole-nesting birds. Assuming that is the case, conservationists have experimented with creating artificial nests designed to suit the needs of quetzals. While some of the nests are essentially wooden boxes, the seemingly best design employs a short section of a hollow decaying log with wood soft enough to be dug out by the quetzals (a key element for success), and a nest hole of the right size for the birds to enter. This is then strapped to a tree in prime habitat. By mimicking what newly paired Resplendents are looking for, these "nests" will hopefully then lure in the birds.

Creating artificial nests still involves art as much as it does science, however, and nest-hunting quetzals often prove to be fussy, ignoring many artificial nests. Studies on nest box design, and on the willingness of quetzals to accept such nests, are still in their infancy, and almost nothing on this subject has been published to date. But this technique, increasingly refined, may soon allow us to boost quetzal breeding numbers.

A nest box for Resplendent Quetzals crafted from a natural log. While Resplendents can be fussy in adopting nest boxes, those built from logs are often the most successful in attracting the birds.

Habitat Protection

Given the state of the natural world, we can cast theory aside and say that it is always a good thing to protect habitat, whether that of quetzals or any other creature, though choices must be made. For quetzals, their nesting grounds, and the forests they visit in the off-season, are clearly important, as are the forest corridors linking both regions. Another boost to quetzal well-being is to plant food trees in marginal habitat. As reforestation efforts accelerate in Mesoamerican highlands, nurseries supplying stock for such planting could prioritize the lauraceous trees whose fruits the quetzals favor. A scattering of such trees in a few dozen key locations, selecting different species with different fruiting schedules, could go a long way toward easing the constraints these birds face as they wander outside the breeding season, and it would also attract the quetzals to spots where people can see and appreciate them.

Such opportunities are important. Getting local buy-in on reforestation efforts is critical. Money channeled through regional conservation groups, and dispersed to landowners in regions where quetzals live, is key to protecting habitat. If field and forest owners begin to concern themselves with the plight of quetzals, not just as beautiful neighbors but as a source of tourist revenue too, these birds have a better chance of becoming long-term survivors in the region.

KABEK community efforts in Costa Rica provide a compelling example of this. Retaining forest, preserving dead trees for nesting, building and siting effective nest boxes, guarding nests from excessive disturbance by tourists and photographers—all these become vital goals for landowners when they realize they can "cultivate" quetzals as part of their landscape and earn good money in doing so.

And we should create incentives for the construction of quetzal lodges, research stations, and visitor centers, the latter functioning to attract and educate tourists, both locals and those from far away.

In my mind, the models for such centers are those that have sprung up around nesting Ospreys (*Pandion haliaetus*), the well-known fish-eating hawk with a global reach. After decades of being shot and trapped by humans, and devastated by pesticides, Ospreys began a remarkable recovery starting in the 1970s and 1980s. And this recovery spurred interest in the species and the development of a handful of Osprey centers in Europe, facilities built near nest sites where people can see and photograph the birds in the wild, and learn about their natural history. One such center built in Scotland, and focused on a single nest, has welcomed more than a million visitors since its inception in the 1970s, generating much needed funds for research while sending tourists home happy to have seen a magnificent bird of prey.

A visitors' center at the Santa Elena Cloud Forest Reserve, Monteverde, Costa Rica. Tens of thousands of tourists visit here each year.

Why not similar centers for quetzals? One could argue that the Cloud Forest Reserve in Monteverde, Costa Rica, approaches this model, although quetzals are not its exclusive focus. There and elsewhere, more could be done to put these birds in focus, and to enhance visitor experiences. Some of these possibilities have already been mentioned: nest cams that bring close-up images of the birds to human visitors; blinds that birders and photographers can rent, carefully placed so as not to jeopardize nesting pairs or feeding groups; and guides and researchers interacting with the public via lectures, video presentations, and writings.

This model should be replicated in other Resplendent Quetzal hot spots such as Chiapas (El Triunfo); Guatemala (Biotopo del Quetzal); Honduras (Santa Bárbara National Park); Panama (Cerro Punta); and Las Tablas and other sites in the southern Talamamacas in Costa Rica. The goal here is to provide not just lodging, meals, and a guide—which some sites are already doing—but to enrich the experience with a center where visitors can gather, learn about the birds, and perhaps have the chance to photograph them more closely without disturbing them.

Right: Coffee farms continue to encroach on the highland forests where quetzals dwell, and this is increasingly the case as warming temperatures compel farmers to move their farms to higher elevations.

We should note that initiatives to involve local citizens, not just foreign tourists, is critical. Current estimates suggest that fewer than one in four locals in Mesoamerica has ever seen a live quetzal; among children under the age of 16, the proportion is even lower. Without knowledge of quetzals, people will never develop pride in their wildlife heritage, which translates into concern for a bird and its habitat. In addition, connecting quetzals to the water that flows out of a tap at home—cloud forests nurturing both—is an educational imperative. The next generation of leaders in quetzal countries need to be aware of such connections. Bringing these birds out of the forest interior and into classrooms, living rooms, and conference rooms is a challenge that clearly needs to be met in the next 20 to 30 years.

Combating Climate Change

While there is much that can be done in the developed world to slow climate change, we have to make an honest assessment of how well those efforts will succeed and how quickly the natural world will change. In Latin American cloud forests, there is no question that warming and drying will accelerate alarmingly quickly over the next half century and beyond. For quetzals, how can we buffer the worst of the projected impacts?

Assuming these birds will move to higher elevations as the forests they live in deteriorate, we could do a few things to help buffer

Bird watchers come from around the globe to see quetzals and cloud forests in Central America. One Japanese tour guide who leads tours to Costa Rica estimates that about 90% of her customers come to Costa Rica with the express goal of seeing the Resplendent Quetzal.

the shocks. Efforts to grow food trees in areas where the birds need them most is an obvious step, one that could make a big difference. In addition, we should consider the possibility of physically moving the birds to safer habitat. Translocations—focused on younger birds, newly fledged—could help these birds colonize higher elevations effectively. Resplendent Quetzals seem to be poor dispersers; tracking data show them returning to familiar nesting areas from one year to the next. But young quetzals likely travel farther before settling down to breed, as is typical of other birds. Efforts to establish (often restore) other avian species in new habitats tend to be most successful when working with individuals at the fledging or post-fledging stage.

So as climate change forces quetzals out of mid-elevation forests, where most individuals live, we should consider moving newly fledged young to higher elevations. Fostering these new populations by enhancing their food supply (planting appropriate trees) and nest sites (crafting nest boxes that succeed in attracting breeders) could help establish the birds. While we cannot be sure if these efforts will be effective, the alternative—losing a large percentage of each quetzal population—suggests that erring on the side of doing more, not less, will be the prudent choice.

The Once and Future Quetzal

Although quetzals—at least the Resplendent—have generated interest among researchers, and extraordinary interest among tourists, there is much more that we could be doing with these birds. We need more people involved, a lot more, not just as researchers but as tourists, photographers, guides, lodge owners, and farmers cultivating quetzal habitat. We see hints of how to achieve that goal—it is already happening in a few areas—but it will take cash, time, and initiative to make it happen more broadly. We can celebrate the fact that we have thriving quetzal lodges in Costa Rica and Guatemala and Panama, but we need to see more such lodges in those countries, in new and different areas, and we need them replicated in countries like Mexico and Honduras, and along the Andes. Having such businesses jumpstart quetzal tourist and research centers would help ensure a bright future for these birds.

Why? In part because we need to generate interest in other quetzal species, not just the Resplendent. And we need to protect cloud forests, wherever they occur and regardless of what quetzal species reside there. In each of the quetzals, evolution has bequeathed us extraordinarily captivating creatures, sustained by the fruits of cloud forest trees, conceived in exuberant flights and extravagant displays, and nurtured as eggs and nestlings in decaying trees. To spend our lives idly without giving these birds their due would be unforgivable. Let's continue to celebrate quetzals, and the forests that nurture them—and our planet that has brought all this to life.

The bird that launched a thousand trips. Far more than that, actually, as more than 100,000 visitors come to quetzal cloud forests each year in Costa Rica alone.

COUNTING QUETZALS

If we were to reinvent the labors of Hercules, we'd be justified in adding something new to his 12 labors, that of completing accurate counts of quetzals. On a level of difficulty, such a count might compare with Hercules having to clean the fabled Augean stables in a day, but instead of having to empty hundreds of cattle stalls of straw and manure, the quetzal challenge would be to fill hundreds of data compartments with numbers that, we hope, are not "manure," and that instead reflect, with a modicum of accuracy, just how many quetzals are actually out there living in Latin American forests, and how much forest remains for them.

Indeed, the challenge is twofold, counting bird populations and estimating the extent of remaining habitat. Such measures are almost entirely lacking for quetzals, even for the relatively well-studied Resplendent, although we have a small handle on numbers for that bird. What's more, methods differ among biologists about how to gather such data. The amount of area to cover in a census and the amount of time spent on doing the census vary from study to study. Some census takers have more finely tuned hearing than others (in thick forest the birds are more often heard than seen). And then one must decide on whether to focus on individual birds or on nesting pairs. All in all, such variation makes it a challenge to develop what biologists call "repeatability," that is, the likelihood that someone who comes after you, using the same methods, will get the same data.

The second challenge is equally difficult, maybe more so. Once we have a measure of quetzal numbers in a known area, then we need to find out how much suitable forest is available to the species so that we can extrapolate to region, country, and even to the entire range of the species. All this assumes that a particular species is living at roughly the same density across its range, at both higher and lower elevations, and in different forests, which is almost certainly not the case. Forest composition, microclimate, fruit availability, predation pressure, nest site availability—all are likely to vary region to region, with quetzal numbers reflecting that. But for our exercise here we assume that, averaged over hundreds of thousands of hectares, such differences even out. We may be wrong in taking that plunge, but we really have no choice if we want to get anywhere near to a population estimate at the end of our exercise.

Estimating forest size remains a particularly elusive goal. Even with the help of satellites that map the earth's surface, and do so with growing sophistication (distinguishing forest types and getting a measure of how intact a forest is), it is still difficult to achieve accurate measures of how much "quetzal forest" remains in Guatemala, for example, or how much "cloud" forest in Columbia constitutes habitat that White-tipped Quetzals could be living in. One of the biggest barriers is simply the vastness of the data to be analyzed. Sorting through reams of satellite data to determine how much forest remains between 1500 and 3000 meters in Costa Rica (the rough altitudinal nesting range of that country's Resplendent Quetzals) is a Herculean task, even with the best of today's computers. But that's a number that would help us refine our estimates of population size in a key part of that species' range.

In addition, nailing down with any precision the range of a species, and how that shifts with geography, remains an elusive goal. Are Resplendents in the Talamancas found at the same elevational range as those in Chiapas, or Monteverde, or Honduras (almost certainly not)? Ditto for Crested Quetzals in Venezuela vs. Ecuador, and do either of those populations migrate to lower altitudes outside of the breeding season (very likely, at least in most years)? We have vague ideas on this for Resplendents, but very little for Cresteds, or for that matter any of the other quetzals. Because of all this, we must admit to our limitations; given what we know today, trying to develop population estimates for any quetzal other than the Resplendent is futile; we have no reliable surveys of those birds, no data on the population density of any of the South American quetzals. Without such numbers there's simply no way to calculate population size.

That said, we can take a stab at estimating Resplendent Quetzal numbers, basing our efforts on a handful of surveys in various parts of the range of that species (most of those surveys conducted by the indefatigable Sofía Solórzano in the early 2000s) and on fairly rough estimates of the size of forested areas in Mexico and Central America that are known to support the species.

Mexico: Surveys by Solórzano (meticulously done, repeated over several years, conducted in different seasons, and counting birds both seen and heard) suggest high quetzal density there, with about 50 pairs/100 hectares in El Triunfo, a key protected area for the species. At least 5000–7000 hectares of quetzal cloud forest remain in that park, with perhaps another 2000–3000 in La Sepultura to the north. Add another 2000 hectares for remaining remnants in nearby areas, and there's likely about 10,000 hectares of forest available to Resplendents in Chiapas, the only region in which this quetzal is found in Mexico. Assuming roughly similar quetzal density throughout these forests would result in about 5000 pairs for Chiapas.

Guatemala: Very few surveys have been conducted there, and none of them are recent, but several by Solorzano in the Sierra de las Minas Biosphere Reserve (SdlMBR) in the early 2000s suggest high breeding density in that key quetzal habitat, with 50 pairs/100 hectares. Estimates suggest that about 120,000 hectares of forest remain in that reserve that could support Resplendents; thus about 60,000 pairs for SdlMBR. For the rest of the country, data are harder to come by, but a rough estimate suggests lower densities of about 30 pairs/100 hectares, and perhaps 50,000 hectares available to the species, giving us about 15,000 pairs. Together, these data suggest a total of about 75,000 pairs for Guatemala.

Honduras: There are almost no survey data for this country, but densities are likely lower than in other areas (owing to forest degradation), with perhaps 20–30 pairs/100 hectares. Satellite data suggest that forest size in the 25 largest and most intact cloud forest reserves in the country is about 295,000 hectares, giving us roughly 60,000 Resplendent pairs for the country. This may be an underestimate; densities could be higher in some forests. Let's hope future surveys will give us more precise numbers.

El Salvador: Surveys by Solorzano suggest low densities here, with about 20 pairs/100 hectares. Very little forest remains, with the largest remnant in Montecristo (Trifinio) National Park (shared with Guatemala and Honduras). The combined area for the park is about 22,000 hectares, with about one-third of that in El Salvador. This gives us about 1500 Resplendent pairs for El Salvador.

Nicaragua: There are no quetzal surveys for the country, but one assumes low densities with perhaps 20 pairs/100 hectares. Although Boswas Biosphere Reserve is immense, only a tiny portion of it supports quetzal cloud forest (about 25,000 hectares), giving us perhaps 5000 pairs. Add another 1000 pairs for other smaller forest reserves in the country, and Nicaraugua might still hold 6000 pairs of Resplendents.

Costa Rica: Recent surveys by myself and others suggest that much of the forest here supports high quetzal densities, with an average of about 50 pairs/100 hectares (this number is probably higher in the southern Talamanacas). Mapping research by F. Joyce and others peg the extent of forest at 1500 to 3000 m (key altitudinal range for the breeding of Resplendent Quetzals in Costa Rica) at about 350,000 hectares in protected areas (parks, reserves). Outside of protected areas, such estimates suggest another 40,000 hectares of forest at that altitudinal range. This results in a combined total of 390,000 hectares of forest likely to support Resplendents in the country. Given those data, there would be about 195,000 pairs in Costa Rica.

Panama: Surveys by Solorzano in the early 2000s suggest densities of about 50 pairs/100 hectares. Core habitat remains so there is little reason to suspect those numbers have changed. Such habitat is centered in Panama's portion of Amistad International Park (about 100,000 hectares of evergreen cloud forest there to support quetzals). Thus the estimate for Panama is 50,000 pairs.

Totals:

> *P. m. moccino*: 140,000 pairs
> *P. m. costaricensis*: 245,000 pairs

This gives us a total of about 385,000 pairs, which in turn is likely to give us close to a million individuals when taking into account non-breeding individuals (especially young birds). Biology suggests this latter group should make up at least 20% of the population.

SUGGESTED READING

What Is a Quetzal?

Cody, S., J. E. Richardson, V. Rull, C. Ellis, and R. T. Pennington. 2010. The Great American biotic Interchange revisited. Ecogeography 33: 326-332.

Collar, N. 2001. Family Trogonidae. Pp. 80-127 in Handbook of the Birds of the World, Vol. 6 (J. del Hoyo et al., eds.). Lynx Edicions.

Dayer, A. A. (2020). Resplendent Quetzal (*Pharomachrus mocinno*), version 1.0. In Birds of the World (T. S. Schulenberg, Editor). Cornell Lab of Ornithology, Ithaca, NY, USA. https://doi.org/10.2173/bow.resque1.01

Eisermann, K. 2012. Noteworthy nesting record and unusual bill coloration of Resplendent Quetzal *Pharomachrus mocinno*. Cotinga 35: 74-78.

Eisermann, K., C. Avendaño & P. Tanimoto. 2013. Birds of the Cerro El Amay Important Bird Area, Quiche, Guatemala. Cotinga 35: 81-93.

Forshaw, J. M and A. E. Gilbert. 2007. Trogons: a natural history of the Trogonidae. Princeton Univ. Press. 304 pp.

Futuyma, D. J. 2021. How birds evolve: what science reveals about their origin, lives, and diversity. Princeton Univ. Press. 320 pp.

Maslow, J. E. 1986. Bird of life, bird of death: a naturalist's journey through a land of political turmoil. Simon & Schuster. 249 pp.

Oliveros, C. H., M. J. Andersen, P. A. Hosner, W. M. Mauck III, F. H. Sheldon, J. Cracraft, R. G. Moyle. 2020. Rapid Laurasian diversification of a pantropical bird family during the Oligocene-Miocene transition. Ibis 162(1): 137-152.

Schulz, U. and K. Eisermann. 2017. Morphometric differentiation between subspecies of Resplendent Quetzal (*Pharomachrus mocinno mocinno* and *P. m. costaricensis*) based on male uppertail-coverts. Bull. B.O.C. 137(4): 287-291.

Simon, H. T. 1971. The splendor of iridescence: structural colors in the animal world. Dodd, Mead and Co., New York.

Solórzano, S. and K. Oyama. 2010. Morphometric and molecular differentiation between quetzal subspecies of *Pharomachrus mocinno* (Troginiformes: Trogonidae). Rev. Biol. Trop. 58(1): 357-371.

Solórzano, S., A. J. Baker, and K. Oyama. 2004. Conservation priorities for resplendent quetzals based on analysis of mitochondrial DNA control-region sequences. Condor 106: 449-456.

Weir, J. T., E. Bermingham, D. Schluter. 2009. The Great American Biotic Interchange in birds. PNAS 106 (51): 21737-21742.

Quetzals in the Mayan and Aztec Civilizations

Berdan, F. F. 1992. Economic dimensions of precious metals, stones, and feathers: the Aztec state society. Estudios de Cultura ~ Vahuatl 2,2: 291-323.

Coe, M. D. and S. D. Houston. 2015. The Maya (9th ed.). Thames and Hudson, 320 pp.

Florescano, E. 1999. The myth of Quetzalcoatl. Johns Hopkins Univ. Press, Baltimore. 150 pp.

Peterson, A. A. and A. T. Peterson. 1992. Aztec exploitation of cloud forests: tributes of liquidambar resin and quetzal feathers. Global Ecology and Biogeography letters: Vol. 2, No. 5: 165-173.

Russo, A. 2002. Plumes of sacrifice: transformations in 16th century Mexican feather art. RES: Anthropology and Aesthetics 42: 226-250.

Schultz, U. and U. Thiemer-Sachse. 2021. La vida del Quetzal en la history del arte mesoamericano. Editorial Idomas, S.L. Unipersonal. Eberswalde Univ. for Sustainable Development.

Sharpe, A. 2014. A Reexamination of the birds in the Central Mexican codices. Ancient Mesoamerica 25(2): 317-336.

Taube, K. 1993. Aztec and Maya Myths (The Legendary Past series). Univ. TX Press, Austin. 80 pp.

Tremain, C. G. 2016. Birds of a feather: Exploring the acquisition of Resplendent Quetzal (*Pharomachrus mocinno*) tail coverts in pre-Columbian Mesoamerica. Human Ecology 44(4): 399-408.

Quetzal Forests

Avila, M. L., Hernández, V. H., and Velarde, E. 1996. The diet of the Resplendent Quetzal (*Pharomachrus mocinno mocinno*) in a Mexican cloud forest. Biotropica 28(4b): 720-727.

Bustamante, M. 2012. Relación de la disponibilidad de frutos de las plantas nutricias del quetzal (*Pharomachrus mocinno*

mocinno de la Llave) con los movimientos altitudinales de quetzales en el gradiente de elevación del Biotopo del Quetzal. Tesis de Licenciatura. Universidad de San Carlos de Guatemala, Guatemala Ciudad.

Hilty, S. 2010. Birds of tropical America: a watchers introduction to behavior, breeding, and diversity. Univ. Texas Press. 312 pp.

Janzen, D. (ed.). 1983. Costa Rican Natural History. Univ. Chicago Press, 815 pp.

Kricher, J. 1997. A Neotropical Companion. Princeton Univ. Press. 451 pp.

LaBastille, B. A. and D. G. Allen. 1969. Biology and conservation of the quetzal. Biol. Conserv. 1: 297-306.

Levey, D. J., C. Martínez del Rio. 2001. It takes guts (and more) to eat fruit: lessons from avian nutritional ecology. Auk 118(4): 819-831.

Nadkarni, N. M. and N. Wheelwright. 2014. Monteverde: Ecology and Conservation of a Tropical Cloud Forest - 2014, Updated Chapters. Bowdoin Digital Commons.

Paiz, M.-C. 1996. Migraciones estacionales del Quetal (/ *Pharomachrus mocinno/* de la Llave) en la region de la Sierra de las Minas, Guatemala y sus implicaciones para la conservación de la especie. Tesis de Licenciatura. Universidad del Valle de Guatemala, Guatemala Ciudad.

Remsen, J. V., M. A. Hyde, and A. Chapman. 1993. The diets of Neotropical trogons, motmots, barbets, and toucans. Condor 95: 178-192.

Skutch, A. F. 1944. Life history of the quetzal. Condor 46: 213-235.

Solórzano, S., S. Castillo, T. Valverde, and L. Avila. 2000. Quetzal abundance in relation to fruit availability in a cloud forest in southeastern Mexico. Biotropica 32(3): 523-532.

Wheelwright, N. T. 1983. Fruits and the ecology of the Resplendent Quetzal. Auk 100: 286-301

Wheelwright, N. T. 1985. Fruit-size, gape width, and the diets of fruit-eating birds. Ecology 66(3): 808-818.

Wheelwright, N. T. 1986. A seven-year study of individual variation in fruit production in tropical bird-dispersed tree species in the family Lauraceae. Pp. 19-35 in Frugivores and seed dispersal (A. Estrada and T. H. Fleming, eds.). W. Junk Publishers, Dordrecht, Netherlands.

Zuchowski, W. 2007. Tropical plants of Costa Rica: a guide to native and exotic flora. Zona Tropical Publications. 529 pp.

At the Nest

Cornelius, C. et al. 2008. Cavity-nesting birds in Neotropical forests: Cavities as a potentially limiting resource. Ornitologia Neotropical 19 (Suppl.): 253-268.

LaBastille, B. A. and D. G. Allen. 1969. Biology and conservation of the quetzal. Biol. Conserv. 1: 297-306.

Skutch, A. F. 1944. Life history of the quetzal. Condor 46: 213-235.

Wheelwright, N. T. 1983. Fruits and the ecology of the Resplendent Quetzal. Auk 100: 286-301.

The Once and Future Quetzal

Blackman, A., A. Pfaff, J. Robalino. 2015. Paper park performance: Mexico's natural protected areas in the 1990s. Global Environmental Change 31: 50-61.

Bubb, P., I. May, L. Miles, J. Sayer. 2004. Cloud Forest Agenda. UNEP-WCMC, Cambridge, UK. Online at: http://www.unep.wcmc.org/resources/publications/UNEP_WCMC_bio_series/20.htm

Bullock, E. L., C. Nolte, A. L. Reboredo-Segovia, and C. E. Woodcock. 2020. Ongoing forest disturbance in Guatemala's protected areas. Remote Sens. Ecol. Conserv. 6(2): 141-152.

Cusack, D. and L. Dixon. 2008. Community-based ecotourism and sustainability: cases in Bocas del Toro Province, Panama and Talamanca, Costa Rica. J. Sustainable Forestry 22 (1-2): 157-182.

Duffy, S. B., M. S. Corson, and W. E. Grant. 2001. Simulating land-use decisions in the La Amistad Biosphere Reserve buffer zone in Costa Rica and Panama. Ecol. Modeling 140: 9-29.

Fadrique, B. et al. 2018. Widespread but heterogenous responses of Andean forests to climate change. Nature 564: 207-213.

Feeley, K. J. et al. 2013. Compositional shifts in Costa Rican forests due to climate-driven species migrations. Global Change Biology 19: 3472-3480.

Gasner, M. R., J. E. Jankowski, A. Ciecka, K. O. Kyle, K. N. Rabenold. 2010. Projecting the local impacts of climate change on a Central American montane avian community. Biol. Conserv. 143: 1250-1258.

Guindon, Carlos F. 1986. The importance of forest fragments to the maintenance of regional biodiversity in Costa Rica. Pp. 168-186 in Forest Patches in Tropical Landscapes (J. Schelhas and R. Greenberg, eds.). Island Press, Wash. DC.

Holl, K. D., M. Loik, E. H. V. Lin, I. Samuels. 2000. Tropical montane forest restoration in Costa Rica: overcoming barriers to dispersal and establishment. Restoration Ecology 8(4): 339-349.

Kappelle, M. 2016. Montane cloud forests of the Cordillera de Talamanca. Pp. 451-491. In: M. Kappelle, ed. Costa Rican Ecosystems. Univ. Chicago Press. Chicago, IL.

Karger, D. N., M. Kessler, M. Lehnert, and W. Jetz. 2021. Limited protection and ongoing loss of tropical cloud forest biodiversity and ecosystems worldwide. Nature Ecology & Evolution 5: 854-862.

Karmalkar, A. V. et al. 2008. Climate change scenario for Costa Rican montane forests. Geophysical Research Letters 35 (11). https://agupubs.onlinelibrary.wiley.com/doi/full/10.1029/2008GL033940

Kockelman, P. 2016. The chicken and the quetzal. Duke Univ. Press, Durham and London. 208 pp.

LaBastille, B. A. and D. G. Allen. 1969. Biology and conservation of the quetzal. Biol. Conserv. 1: 297-306.

Martin, T. E. and G. A. Blackburn. 2009. The effectiveness of a Mesoamerican "paper park" in conserving cloud forest avifauna. Biodivers, Conserv. 18: 3841-3859.

Molina Murillo, S. A., J. P. Pérez Castillo, M. E. Herrera Ugalde. 2014. Assessment of environmental payments on indigenous territories: the case of Cabecar-Talamanca, Costa Rica. Ecosystem Services 8: 35-43.

Nadkarni, N. M. and N. Wheelwright. 2014. Monteverde: Ecology and Conservation of a Tropical Cloud Forest - 2014, Updated Chapters. Bowdoin Digital Commons.

Nagendra, H., C. Tucker, L. Carlson, et al. 2004. Monitoring parks through remote sensing: studies in Nepal and Honduras. Environmental Management 34(5): 748-760.

Pfeffer, M. J., J. W. Schlelhas, S. D. DeGloria. 2005. Population, conservation, and land use change in Honduras. Agriculture, Ecosystems, and Environment 110: 14-28.

Ponce-Reyes, R., V.-H. Reynoso-Rosales, J. Watson, J. VanDerWal, R. A. Fuller, R. L. Pressey, and H. Possingham. 2012. Vulnerability of cloud forest reserves in Mexico to climate change. Nature Climate Change 18 Mar 2012; DOI: 10.1038/NCLIMATE1453

Pope, I., D. Bowenc, J. Harbora, G. Shaod, L. Zanottie, G. Burniske. 2015. Deforestation of montane cloud forest in the central highlands of Guatemala: contributing factors and implications for sustainability in Q'eqchi' communities. International J. of Sustainable Development & World Ecology 22: 201-212.

Pounds, J. A., M. Fogden, and J. H. Campbell. 1999. Biological response to climate change on a tropical mountain. Nature 398: 611-615.

Powell, G. V. N. and R. D. Bjork. 1994. Implications of altitudinal migration for conservation strategies to protect tropical biodiversity: a case study of the Resplendent Quetzal (Pharomachrus mocinno) at Monteverde, Costa Rica. Bird Conservation International 4: 161-174.

Powell, G. V. N. and R. D. Bjork. 1995. Implications of intra-tropical migration on reserve design: a case study using Pharomachrus mocinno. Conservation Biology 9(2): 354-362.

Rehm, E. M. and K. J. Feeley. 2015. The inability of tropical forest species to invade grasslands above treeline during climate change: potential explanations and consequences. Ecography 38: 1167-1175.

Renner, S. C. and M. Markussen. 2005. Human impact on bird diversity and community structure in a tropical montane cloud forest in Alta Verapaz, Guatemala, with special reference to the Quetzal (Pharomachrus mocinno). In: Valuation and Conservation of Biodiversity (M. Markussen et al., eds). Springer-Verlag Berlin Heidelberg.

Renner, S. C., M. Voigt, and M. Markussen. 2006. Regional deforestation in a tropical montane cloud forest in Alta-Verapaz, Guatemala. Ecotropica 12: 43-49.

Rodriguez Zuniga, J. M. 2003. Paying for forest environment services: the Costa Rican experience. Unasylva 212: 31-33.

Solórzano, S., M. A. Castillo-Santiago, D. A. Navarrete-Gutiérrez, and K. Oyama. 2003. Impacts of the loss of neotropical highland forests on species distribution: a case study using the resplendent quetzal, an endangered bird. Biol. Conserv. 114: 341-349.

Solórzano, S., A. J. Baker, and K. Oyama. 2004. Conservation priorities for resplendent quetzals based on analysis of mitochondrial DNA control-region sequences. Condor 106: 449-456.

Toledo-Aceves, T., J. A. Meaves, M. González-Espinosa, N. Ramírez-Marcial. 2011. Tropical montane cloud-forests: current threats and opportunities for their conservation and sustainable management in Mexico. J. Enviro. Management 92: 974-981.

Tosi, J. A. J., V. Watson, and J. Echeverria. 1992. Potential impacts of climate change on the productive capacity of Costa Rican forests: a case study. Tropical Science Center, unpubl.).

Walker, K. 2020. Capturing ephemeral forest dynamics with hybrid time series and composite mapping in the Republic of Panama. International J. of Appl. Earth Observation and Geo-information 87, May 2020, 102029.

Wallace, D. R. 1992. The Quetzal and the Macaw: the story of Costa Rica's National Parks. Sierra Club Books, San Francisco. 222 pp.

IMAGE CREDITS

1300 I St. Northwest Suite 400E - #9996
Washington, DC 20005
202-765-2266
www.osaconservation.org

CONSERVING COSTA RICA'S NATURAL TREASURE

We Need The Wild
An Overview of Osa Conservation

Situated on the southern pacific coast of Costa Rica lies one of the most biodiverse places on Earth. A land mass smaller than Los Angeles county, the Osa Peninsula boasts exceptional levels of endemicity and rarity with over 400 species of birds, 140 mammals, 500 trees and 6,000 insect species described so far. Endangered spider monkeys hang from ancient megatrees, jaguars track giant herds of white-lipped peccaries, and tapirs crunch through the largest lowland tropical rainforest on Central America's pacific slope.

The incredible and complex ecosystems found here provide invaluable services to the people who depend on them for clean air and drinking water, jobs, cultural resources, and a stable climate—and so their conservation in perpetuity is critical.

We believe successful conservation must be a locally supported endeavor with international implications. In this sense, we aim to replicate the "small but mighty" brand that distinguishes Costa Rica as a world environmental leader. Costa Rica is a green giant where unparalleled strides in conservation have propelled this small country to the national stage. With unprecedented recovery from aggressive extraction of the mid-1900s, over the past few decades, Costa Ricans have doubled down on defending nature: investing in education instead of the military, paying landowners to conserve and restore forests, and giving lawful protection to riparian forests instead of clearing them. Most importantly, they worked to establish a network of protected areas throughout the country. These pioneering efforts mean that forest cover has increased by over 150% since the 1980s.

Our approach is place-based. We are rooted in the biological and social landscape. We use a scientific, evidence-based approach to conservation, and invest in learning, education, and employing knowledge to solve environmental problems. Adaptation to change and innovation are values we pursue and encourage – we aspire to serve as a global model for excellence in conservation. Our impact stretches beyond Osa lowlands, to the cloud-shrouded elfin forests of the Talamanca mountains. This is the same elevational landscape used by elevational migratory birds, such as the resplendent quetzal, and the three-wattled bell-bird. We are building a working model of climate adaptation to protect biodiversity and build resilience for both people and nature.

Osa Conservation was founded in 2003 with the mission to protect the incredible biodiversity of the Osa Peninsula, in full commitment to improving lives through conserving nature.